The New Engineering Research Centers
Purposes, Goals, and Expectations

Cross-Disciplinary Engineering Research Committee
Commission on Engineering and Technical Systems
National Research Council

NATIONAL ACADEMY PRESS
Washington, D.C.   1986

NATIONAL ACADEMY PRESS 2101 Constitution Ave., NW Washington, DC 20418

NOTICE: The project that is the subject of this report was approved by the Governing Board of the National Research Council, whose members are drawn from the councils of the National Academy of Sciences, the National Academy of Engineering, and the Institute of Medicine. The members of the committee responsible for the report were chosen for their special competences and with regard for appropriate balance.

This report has been reviewed by a group other than the authors according to procedures approved by a Report Review Committee consisting of members of the National Academy of Sciences, the National Academy of Engineering, and the Institute of Medicine.

The National Research Council was established by the National Academy of Sciences in 1916 to associate the broad community of science and technology with the Academy's purposes of furthering knowledge and of advising the federal government. The Council operates in accordance with general policies determined by the Academy under the authority of its congressional charter of 1863, which establishes the Academy as a private, nonprofit, self-governing membership corporation. The Council has become the principal operating agency of both the National Academy of Sciences and the National Academy of Engineering in the conduct of their services to the government, the public, and the scientific and engineering communities. It is administered jointly by both Academies and the Institute of Medicine. The National Academy of Engineering and the Institute of Medicine were established in 1964 and 1970, respectively, under the charter of the National Academy of Sciences.

This activity was supported by the National Science Foundation under cooperative agreement No. ENG-8505051 between the Foundation and the National Academy of Sciences. The opinions, findings, and conclusions or recommendations are those of the committee and the speakers and do not necessarily reflect the views of the National Science Foundation.

LIBRARY OF CONGRESS CATALOG CARD NUMBER 85-51808

INTERNATIONAL STANDARD BOOK NUMBER 0-309-03598-8

Printed in the United States of America

CROSS-DISCIPLINARY ENGINEERING RESEARCH COMMITTEE

SEYMOUR L. BLUM, Vice-President, Charles River Associates, Inc. (*Chairman*)
ROBERT R. FOSSUM, Dean, School of Engineering and Applied Science, Southern Methodist University
JAMES F. LARDNER, Vice President, Component Group, Deere & Company

Staff

KERSTIN B. POLLACK, *Executive Secretary*
COURTLAND S. LEWIS, *Consultant*
VERNA J. BOWEN, *Administrative Assistant*
DELPHINE D. GLAZE, *Administrative Secretary*
PATRICIA WHOLEY, *Fiscal Assistant*

Preface

In the fall of 1983 a small group of engineers met with George Keyworth II, the President's Science Adviser and Director of the Office of Science and Technology Policy. The stated purpose of the meeting was to present to Dr. Keyworth a briefing on the need for advances in research on the use of computers in design and manufacturing. The briefing had been prepared under the auspices of a National Research Council committee, of which the late George Low was chairman.

As the meeting progressed, however, its focus shifted from the overt subject of computer-aided design and computer-aided manufacturing (CAD/CAM) to a single theme that lay in the background, hidden between the lines of the report. That underlying theme was the need for integration of the engineering endeavor. It was a subject that had cropped up here and there, more and more often over the previous three or four years, in studies and pronouncements about engineering research and education. The topic was usually alluded to as though in passing, with an air that "this is important, but hard to grasp." References to it were especially frequent whenever concerns about our declining overall competitiveness in technology-intensive, manufacturing-oriented industries were being discussed.

The need for integration has many facets, and can be expressed in many ways: the integration of engineering research and development, of design and manufacturing; the closer interplay of universities and industry; the greater exposure of engineering students to practical, hands-on, apprenticeship aspects of education. One particularly important element identified is the need for a new, crosscutting approach to complex engineering research problems—often expressed by the key words cross-disciplinary, interdisciplinary, and multidisciplinary. The traditional disciplines alone are not always suited to the complex nature of modern engineering discovery. Systems is another key term, referring not just to systems engineering, but to the need for attention to the systems aspects of the engineering enterprise and its products, and for optimizing the overall process by considering every element, looking for trade-offs, incorporating diverse kinds of expertise, taking the broadest possible view.

These concerns had been latent—there, but not addressed. For one thing, they were elusive, hard to define. There was no real knowledge base to support any rigorous discussion or definition of the problems or, for that matter, what was at stake. The ideas seemed likely to challenge the structure and function of the engineering research establishment. Cross-disciplinary research and university-industry partnership were concepts that augured major change without any guarantee of commensurate return. But by the fall of 1983 the undercurrent of interest in this theme had reached a point of critical mass in the minds of those concerned with the nation's technological competitiveness.

So it was that the group in Dr. Keyworth's office began to discuss these ideas with a sense of growing excitement. Dr. Low in particular catalyzed a shared vision of the kind of engineering education that is needed if this kind of integration were to be achieved in the universities and in industry. A second meeting was arranged to discuss what might be done. As a result of that meeting, the National Science Foundation (NSF) became involved with a new agenda to create university-based cross-disciplinary research centers that would be closely attuned to the perceived real engineering needs of the nation.

In December 1983 the NSF asked the National Academy of Engineering to conduct a brief study of the engineering research center concept, aimed at formulating guidelines for the centers' mission, organization, operation, and funding. The results of that study were transmitted in February 1984, and by April 1984 the first NSF program announcement for the Engineering Research Centers (ERCs) was issued.

The response was enormous: 142 proposals were received from more than 100 universities for research in a wide range of fields. After an exhaustive review, awards for six Engineering Research Centers (involving a total of 8 universities) were announced in early April 1985. The papers presented here were delivered at a symposium held later that month at the National Academy of Sciences to introduce the new Centers to the engineering community at large.

NSF's expressed purpose in supporting these Centers is to provide cross-disciplinary research opportunities for faculty and students, to provide fundamental knowledge that will contribute to the solution of important national problems, and to prepare engineering graduates who possess the diversity and quality of education needed by U.S. industry. As Dr. Keyworth pointed out in remarks following his speech on the first day of the symposium, "The ERCs are the real gem of all the new programs that are receiving so much emphasis in fundamental research and the training of talent today. This 'institute' concept . . . is something that is long overdue in this country, and I think it is going to become big."

We concur wholeheartedly with that assessment. The ERCs are the right step at the right time; they will inject into engineering new values and new approaches that are sorely needed. It behooves all of those involved in the engineering enterprise in the United States to ensure that this gem is highly polished, and that the sparkle and promise of this new beginning are not permitted to fade.

Symposium Steering Group
SEYMOUR L. BLUM, *Chairman*
ROBERT R. FOSSUM
JAMES F. LARDNER

Contents

IV   The Future—Challenges and Expectations

The New Engineering Research Centers
Purposes, Goals, and Expectations

SYMPOSIUM
Washington, D.C., April 29–30, 1985

Summary

The symposium titled "The Engineering Research Centers: Factors Affecting Their Thrusts" was held on April 29–30, 1985, under the auspices of the National Research Council's Commission on Engineering and Technical Systems (CETS). The two-day event drew more than 400 representatives of academe, industry, and government to hear speakers describe the Engineering Research Centers (ERCs), the concept behind them, and their importance to the nation's future. Discussion was encouraged, so the symposium became the forum for a lively interchange of ideas about the Centers and, indeed, about the present and future status of the engineering research and development enterprise in the United States. (The discussion that followed each presentation is summarized in this symposium volume immediately after each paper.)

The first session opened in the afternoon with a series of presentations describing the national goals that the ERCs represent. George A. Keyworth II, Science Adviser to the President and Director of the Office of Science and Technology Policy, gave the keynote address. He and other leaders in government discussed the relation of the Centers, and of engineering research in general, to international industrial competitiveness. Mr. Erich Bloch, Director of the National Science Foundation (NSF), described the continuity that exists between science, engineering, and technology, and which must be more widely accepted if the nation's economy is to benefit from a strong industrial competitiveness across a broad front.

The next group of speakers spoke of the ERCs from the point of view of the NSF—the concept behind them, their goals, selection criteria applied in the first round of awards, and mechanisms for support of the current and future Centers. Symposium participants representing universities with an interest in hosting ERCs of their own were especially attentive to this portion of the program; they asked a large number of questions relating to the review process for future selection cycles.

The following morning the symposium reconvened to hear presentations by each of the new Center directors of the programs they plan to introduce to meet the Centers' goals. The varied programs for research, education, and industrial liaison were of central interest; these were impressive in scope and in creative initiative. Tentative ideas regarding the establishment of mechanisms for the exchange of information and technology among the ERCs and their respective research communities were also presented. Because the educational function of the Centers is as important as their research function, a view of the relationship between these two functions in the context of modern engineering was expressed by the chairman of a National Research Council committee that had just concluded a study of the subject.

The final session of the symposium entailed a look at the future of U.S. industry and engineering from the standpoint of challenges that will have to be met and expectations that the ERCs will be called on to fulfill. Speakers outlined the roles that the Centers can play in aiding and stimulating mature industries (e.g., the automotive industry), growth industries (e.g., electronics and computers), and emerging industries (such as biotechnology). They stressed that this bold new approach to engineering research and education carries with it a range of new responsibilities not only for academe, but—just as important—for industry and government. Each of these traditionally separate sectors will be challenged to cooperate in the nurturing and support of the ERCs, a fact which the final group of speakers emphasized.

The predominant message that emerged from the symposium is that this is the beginning of a new era, in terms of world technological and economic dynamics and in terms of the roles of engineering practice and research. The ERCs are among the first deliberate responses the nation has made to that changing environment: new engineering institutions designed for the new era. The goal of the program is "to develop fundamental knowledge in engineering fields that will enhance the international competitiveness of U.S. industry and prepare engineers to contribute more effectively through better engineering practice." The explicitly economic and practical nature of that goal is in itself a novel feature, and one that is likely to be seen more and more often in the future.

The 6 ERCs introduced at the symposium are only the first contingent of what the NSF expects eventually to grow to some 20 Centers, each with an average annual budget of $2–$5 million. And, as was noted by both Erich Bloch and George Keyworth, other agencies of government besides the NSF are interested in pursuing the ERC concept. George Keyworth expressed his belief that the ERCs might well come to represent "something on the order of 10 percent of the entire National Science Foundation budget in a very short period of time." Given the likely participation of other agencies, he pointed out, the total number of such centers could very quickly exceed the anticipated 20.

The existence of a large number of Centers, focusing on different areas of engineering research, will require a broad base of support. In the case of the ERCs, NSF support is not envisioned as permanent, but as start-up funding. The awards will be made as continuing grants for an initial duration of five years. During that time the Centers are expected to have established a strong network of relationships with industry, and to have obtained substantial industrial support. In this way, where feasible, the Centers should eventually become self-sufficient, requiring no further NSF support.

Such a goal clearly places several requirements upon the ERCs. First, they must be sure to establish the kind of industrial liaison programs that will lead to continuous and mutually beneficial interactions. The plans and programs described by the six Center directors are a good start in this direction.

Second, the ERCs must produce high-quality research, the results of which are useful to industry without being too near-term in focus. As Roland W. Schmitt characterized the Centers, they will bridge the gap between the generation of knowledge and its application to the market-place. "From industry . . . should flow . . . the barrier problems that practice is running up against. From universities . . . should flow the knowledge and talent needed to overcome the fundamental problems." To that end Susan Hackwood, Director of the new Center for Robotic Systems in Microelectronics at the University of California at Santa Barbara, envisions a procedure in which researchers at the Center will "go from the specific to the general, doing applications first and gaining fundamental knowledge later."

Third, the ERCs must attain self-sufficiency by performing their educational function well. If they can attract top students, both graduate and undergraduate, and inculcate in them a broad understanding of what is needed to bring sophisticated products all the way from the laboratory to market, the graduates of the Centers will become a most effective form of advertisement for the cross-disciplinary ERC approach to research.

What can industry do, for its part, to ensure the success of the ERCs? As James Lardner of Deere and Company puts it, industry must:

- help identify and define manufacturing research needs that offer intellectual challenges to the academic community that are commensurate with established research activities;

• make available selected, experienced industry representatives to support research projects;

• be willing to provide constructive input for program evaluation and to make recommendations to enhance the value of the research findings . . . the Centers produce;

• recognize, hire, and reward the graduates of the Centers, offering opportunities commensurate with their potential.

A basic requirement is for industry to be aware of the activities of the Centers. Participation in the exchange networks described by NSF's Carl Hall would be a simple and effective means to maintain such an awareness. In general, industry managers can help the ERCs attain their goals by being open to the opportunity they represent—that is, by avoiding the pressure for near-term results, by not being restrictive in the approach to joint research and the publication of results, and by taking advantage of the continuing educational possibilities they will afford.

Perhaps the greatest adjustment will be required by universities that host the ERCs. As Semiconductor Research Corporation president Larry Sumney noted, universities are structured around discipline-oriented departments. The cross-disciplinary environment of the Centers runs counter to this traditional structure, and the effect on a faculty member's status and career can be severe if the ERC is not accepted and integrated within the university's culture. H. Guyford Stever emphasized the need for changes in this "campus sociology" if the ERCs are not to be rendered vulnerable. Strong commitment on the part of university administrators, faculty, and graduate students alike will be essential. To achieve that degree of commitment the universities will have to become sensitive to the nation's economic and competitive needs, and recognize that engineering is the key to fulfilling those needs.

Government also has major responsibilities in this regard, as outlined by Nam Suh, Assistant Director for Engineering at the NSF. Apart from its role as the investment organization, or catalyst, the NSF is also the enabling agent that will help the ERCs overcome problems and achieve their goals. It will also be the NSF's responsibility to secure the continuing support of the Congress and other government entities for the ERCs and the concept they represent. In addition, the NSF plans to encourage state governments to provide joint or independent funding for ERCs or similar research organizations. Nam Suh notes, however, that "in the final analysis no government can be greater than the people it represents," so the willing support of the engineering community in academe and industry will be the real key to the continuing support of the ERCs.

What is the likely outcome if the ERCs are successful? What advances or changes are we likely to see in research, in engineering education, and—perhaps most crucial—in the health of our industries? Several speakers gave their views, their visions, of what the results might be. Nam Suh hopes that the ERCs "will come up with concepts and ideas that, 20 years from now, can change the way we live, the way we function, and the way we produce goods." He believes that the ERCs have the potential to create for engineering a climate of discovery similar to that which appeared in physics at universities throughout Western Europe in the early part of this century—an "exciting cultural environment which will create new intellectual frontiers and many important breakthroughs."

The changes in engineering education are likely to be substantial for participating students. Roland Schmitt pointed out that it has been difficult for a student to acquire both the needed scientific knowledge and the apprenticeship aspects of education. Unlike education in the sciences, it is rare for engineering graduate students to be trained in the type of facilities they will encounter in industry. And engineering is the only profession in which teachers are not, by and large, experienced practitioners. Jerrier Haddad believes that the ERCs will go a long way toward changing this situation. For one thing, the closer contact of academic researchers with industry problems and methods will make them better teachers. More fundamentally, however, participation of students in ERC research programs will be a form of interning. It will introduce the missing element of practice, conferring practical values, greater interest in the work, and stronger personal development as well.

Clearly the real focus of the ERC concept, from the standpoint of both research and education, is the improvement of our national industrial competitiveness. If the ERCs can provide a strong link between academe and industry, research and development, education and practice, they can vastly improve the effectiveness with which we apply our rich national resources of knowledge and talent. If they can bridge the traditional engineering disciplines they can be the catalyst for a needed reshaping of research approaches and values, in universities as well as in industrial manufacturing practices. As George Keyworth observed in his keynote address, "This removal of barriers lies at the heart of the new Engineering Research Centers." It will be necessary that everyone—those in academe, in industry, and in government—understand why those barriers must come down, and that all work with a will to help the ERCs succeed.

Introduction

H. GUYFORD STEVER

This symposium marked the beginning of a brave new venture in American technological enterprise. For those who have participated in their making, the Engineering Research Centers have been eagerly awaited. For a few dedicated individuals who long ago saw the need for a new approach to engineering research, education, and practice, this is a venture that has been long in the making.

Some 300 members of the engineering community attended the symposium to share the excitement of the ideas embodied in the Engineering Research Centers (ERCs). In their papers leaders of the business and academic communities and leaders in government describe the difference that this new concept will make, the opportunity that the Centers present. They describe the roots of the ERC concept and program, the effort, energy, and ideas that went into their creation. The directors of the new Centers and others discuss their plans for making the Centers strong and successful. We read of challenges that the future will present to U.S. industry, as well as to the Centers themselves. And we are confronted, in turn, with the challenges that the Centers present to industry, academe, and government if they are to become an effective instrument for keeping the nation technologically strong and vibrant in the uncertain years ahead.

As a broader audience now begins to share in the excitement of this venture we should not lose sight of what we are about. In some ways we are attempting through the ERCs to change the system, to push engineering research and education over a threshold into a new way of doing things. So it is extremely important that we get it right from the beginning, and that our purposes, goals, and expectations with regard to the ERCs be clear. The symposium was indeed a debut, and this volume is its official announcement. I hope that all who read these papers will be charged with hope, eagerness, and a sense of responsibility for the commitment to the success of the Engineering Research Centers which we must all share.

I

The National Goal

Improving the U.S. Position in International Industrial Competitiveness

GEORGE A. KEYWORTH II

People who have heard me speak on the subject of the National Science Foundation's Engineering Research Centers program know how strong my commitment to the concept is, and how much I look forward to the testing of the concept that is beginning now. The people connected with the first six Centers are to be congratulated. The good news is that they have survived what may have been the toughest grant competition in the NSF's history. The bad news is that they now have to do all those things they promised in the proposals. Actually, I would be disappointed if their new experiences didn't force them to diverge from those plans very quickly, because they are traveling where no one has gone before. They are trying to adapt institutions steeped in tradition to rapid changes in the world of science and technology and in the way those changes are transferred to industry. They are going to have to learn—and teach the rest of us—as they progress.

As someone with a deep interest in the Engineering Research Centers (ERCs), I will try to describe the Centers in the broader context of American industrial competitiveness and of the kinds of resources we have to mobilize to be successful. To set the stage, I want to share a recent experience. The occasion was a conference of delegations from two dozen economically advanced nations who were invited to Venice by the Italian prime minister to discuss the relationship between technology and employment. The event was spurred in part by the growing divergence between the economies of Europe and those of countries, like the United States and Japan, that have been aggressive in taking advantage of new technologies. The European nations have struggled just to maintain the same number of jobs for nearly 15 years. During that same time in the United States we have created 26 million new jobs. Not surprisingly, then, most of Europe today is faced with massive unemployment, with problems so severe that some countries now talk about entire generations of young people who will never find jobs.

One would have expected the European nations to be curious, if not eager, to learn from dynamic economies elsewhere. Yet I came away from that conference very disturbed by what I interpreted as an ingrained resistance to change among many of the European leaders who were there. I was amazed at the number of European officials who proposed that the way to create jobs was to shorten the workweek so that four people might be able to do the work of three. That's hardly what I would call innovation. Others insisted that their high priorities were to provide either what they called "humane" employment, accommodating the life-styles to which the workers have become accustomed, or guaranteed financial support for a comfortable life of unemployment. While they all seem to understand the need to use technology to develop new industries and modernize old ones, when it came to considering actions many of them saw technology as a threat rather than an opportunity. In the true "Europessimist" sense, they could see only the possibility of jobs being eliminated by new technology and productivity improvements, never the jobs that would be created. Not surprisingly, one of my favorite words, "competitiveness," rarely crept into the discussion; it was as if competition simply were not an element of the industrial world.

As we know, competitiveness is a key word where economies are growing. One of the points I tried to make at the conference was that neither world nor domestic trade is a zero-sum game. Technological advances, by increasing the productivity of both labor and resources, create and enlarge markets. In other words, it is not simply a matter of cutting the pie differently; technological advances can make the pie larger. To illustrate this point I cited the example of the personal computer. Just four years ago the market for personal computers was still fairly small. Since then IBM has entered the market, and IBM alone will sell almost $7 billion in personal computers worldwide this year. Yet more than half the parts in the IBM PC are manufactured in other countries and imported to the United States. So in spite of how unexceptional those transactions may appear in light of trade balances, all the countries whose industries are involved in the new enterprise benefit from expanded employment.

I may not have made many new friends when I pointed out to the Europeans that it looks odd for them, with their strong industrial, technological, and educational bases, to be wringing their hands in dismay while at the same time newly industrializing nations, especially in the Far East, are building new technology infrastructures from scratch and becoming formidable competitors in carefully chosen niches of the world's industrial market. Considering these emerging industries, such as Korean steel, Taiwanese electronics, and Indonesian aircraft, it is beyond me how already well-established European (or American) industries, with their expertise and experience, can argue that they operate at a competitive disadvantage. This is the argument we would expect from countries trying to break into a market strongly dominated by established industrial nations.

The lesson I would draw from these observations is that the most important determinant of industrial success these days is a willingness to grasp the opportunities offered by changing technology. I would add that even strong national R&D commitments, as necessary as they are, must still be supplemented by competitive spirit.

I would have been even more depressed at the contrast between Europe and United States in 1985 if I had not reminded myself that societies can become energized with a desire to change and to compete. In the United States we have certainly responded positively to the industrial and technological challenges of the past generation. Admittedly, at the start of this decade we suffered some confusion over the nature of our new competition. Our experience of relatively easy market domination in the past had not prepared us for our new role.

This experience, I'm convinced, will also be positive in the long run, because it is forcing us to reexamine and reaffirm the principles of our economy, and it is forcing us to recognize how much we had dulled our initiative by taking our industrial strengths for granted. Today we not only have a more realistic view of our competition; we also have a more realistic view of our significant capacity to compete. To the extent that one can characterize a national mood, I would say that the American people and American industry are more optimistic today than they've been in years, and that they are looking forward to a healthy economic future.

One example is worth sharing. In March 1985, at a small lunch that President Reagan had with some leaders of American high technology, one of the guests reached into his pocket and pulled out a wafer just off a new manufacturing production line for 1-megabit RAM chips. In displaying the chip this guest was making two points. First, he reminded us that only four years ago many people were ready to dismiss American manufacturing of RAM chips because the Japanese had presumably captured the future markets with their then-advanced 64K RAMs. The guest wanted to remind us that listening to pessimists can be very bad business practice. Fortunately, his company and others had confidence in their abilities and, clearly, had bounced back.

This man was also pointing out the tremendous rate of growth in one particular kind of microelectronics technology. In less than a decade we went from 2 kilobits to 1 megabit. The 4-megabit chip isn't far over the horizon, and I expect to see a 64-megabit chip within my own working lifetime. However, I don't think there is anyone who knows how we are going to use memory devices of that incredible capacity. In fact, the big chips that industry is producing are already stimulating us to rethink the ways we process and use information, leading us right back to basic research. As a result of these industrial advances, we are now investigating entirely new kinds of computing and data-processing technologies. Academic researchers are already beginning to explore the new computer architectures, software, and mathematics that these industrial advances point to. Today's computer, which has been evolving for four decades, may become a thing of the past. Meanwhile, the rate of change in these areas is breaking down traditional barriers between industry and basic research laboratories—barriers that have impeded progress for too long. This removal of barriers lies at the heart of the new Engineering Research Centers.

A few months ago the President's Commission on Industrial Competitiveness completed its 18-month-long analysis of what we have to do as a nation to enable our industries to compete effectively in world markets. One of the points I found especially interesting was the conclusion by this group, which was composed primarily of industrial leaders, that the United States has only two competitive advantages in today's international market of low-cost labor, overvalued dollars, high interest rates, and byzantine trade regulations. Those two advantages are our scientific and technical knowledge base and our talent base.

While the conclusion that knowledge and talent are important American industrial advantages is hardly surprising, I think that all of us on the Commission were surprised to find that they were of such paramount importance. As a consequence, one of the Commission's major conclusions was to endorse the strong and increasing commitment to R&D over the past five years by both industry and the federal government; in addition, the Commission urged creation of "a solid foundation of science and technology that is relevant to commercial uses."

This sounds very much like the point of the Engineering Research Centers. The ERCs may be a preview of new mechanisms to take advantage of the changing relationships between the laboratory and the factory. Over the next few years the ERCs will be helping us to learn a lot about how to improve something we have never paid too much attention to before: the ways universities and industry can cooperate—not just to speed the flow of new knowledge into applications, although that is a major objective, but also to encourage universities to take advantage of industrial expertise in thinking about academic research directions and educational objectives.

Over the past few years many people have concluded that notwithstanding the remarkable successes of American universities in advancing knowledge in science, their structure is not as well suited to the challenges posed by today's industrial opportunities. The narrow approach to research, in which studies are generally confined to highly specialized subdisciplines, needs to be joined with broader perspectives.

The overwhelming response of the universities themselves to this new program—there were proposals from virtually every engineering and research university in the country—reveals what I can only interpret as tremendous enthusiasm for breaking out of some of the old molds of education and research, an impression intensified by my observation of the many people present at the symposium. The establishment of what are in effect campus institutes where academic and industrial scientists and engineers can work together on the kinds of technical problems now being generated by modern industry may mark a new path for science and engineering education and research. One of the most important products of the ERCs will be the students, who will emerge with the broad technical skills that will be needed in tomorrow's industrial world.

To industrial representatives interested in the Centers I can offer assurances, on behalf of the President and his budget advisers, that they will be welcomed as financial partners in this enterprise. But in all seriousness, what is far more important is the enthusiasm of industrialists, their participation, and their commitment to having an impact on how these Centers evolve.

To appreciate why this is important we should consider the origins of the Centers. The idea surfaced in a presentation to my office on the subject of computers in design and manufacturing, made by the Committee on Science, Engineering, and Public Policy (COSEPUP).* The presentation brought home to all of us how radically the role of the engineer will change in light of the tremendous information-processing capabilities that are emerging, such as that 1-megabit RAM chip. We realized, too, that the example of information technology, while perhaps the best known, was only one of many rapidly changing fields that will change engineering.

After that presentation we were convinced that we should be doing more to help integrate engineering practice and training with these new areas of technology and science, and that our future industrial successes were going to depend on the availability of different kinds of engineers than those who had been successful in the past. We turned to the National Academy of Engineering (NAE), which quickly assembled a group to

---

*COSEPUP is a joint committee of the National Academies of Sciences and Engineering and the Institute of Medicine.

suggest new mechanisms through which the National Science Foundation and universities could respond. In both the COSEPUP panel and the NAE group engineers from industry were full and eager participants. The program that emerged has been strongly influenced by industry, so the Centers should be prepared for fruitful interactions there.

This program is a superb example of what we can do together. Some of the general goals guiding government actions to capitalize on our knowledge and talent can be briefly summarized.

First, over the past four years our government has reversed its priorities in order to support the generation of knowledge and talent, rather than the development of specific technologies. Government does not have the ability to guide the development of competitive new industrial technologies. It simply cannot respond rapidly enough to change. Industry itself is far better prepared to make the necessary decisions, and also to make the necessary investments in new technologies to meet demands. On the other hand, support for basic research and for training students is properly the government's responsibility, because both those efforts build the knowledge and talent base.

In 1981 technology development claimed the largest fraction of U.S. government support for research and development, while support for basic research had the smallest fraction. By 1984 those priorities had been reversed—the result of a nearly 60 percent rise in government funding for basic research from 1981 to 1985. Even though federal budgets have been tightly constrained, we never considered it a luxury to allocate resources to such fields as mathematics, physics, chemistry, engineering, and the biological sciences. These investments in pioneering research will lead to tomorrow's new technologies and to tomorrow's economic strength.

Second, we believe government has a responsibility to help universities create the environment needed to be in the forefront of basic research and the education of new technical talent. Our challenge today, reflected in the new Engineering Research Centers, is to sustain creativity and innovation while reducing the barriers between the pursuit of knowledge and the pursuit of productivity.

One major step we have taken to meet this challenge has been to provide such large increases in government support for basic research in universities. We have also increased funding to replace outdated research equipment, improved the access of university researchers and their students to supercomputers, and, together with industry, created special programs to attract the best young engineers and scientists to teaching and research careers in universities.

I have already discussed government's third major responsibility: finding better ways to stimulate the flow of ideas, expertise, and people among our extensive government research laboratories, the universities, and industry. Arrangements like the ERCs are good examples of how we can do that.

Finally, the fourth goal of government for science and technology is to be more alert to emerging technological opportunities and to make sure that we develop the best knowledge and talent base for industry to draw on. In the past our government has not always paid sufficient attention to the opportunities for doing this, and some opportunities have been lost. Lost opportunities in today's highly competitive world can be very expensive. For example, over the years our federal government has spent billions of dollars on the molecular biology that made possible today's biotechnology industry. But by focusing so intently on medical applications we may be failing to develop similarly far-reaching applications in agriculture, and even in manufacturing. In the United States, as in many other countries, there is a real danger of letting others assume industrial leadership in profitable new fields of technology, even though we have a head start through immense investments in the research which has established those fields.

Returning to my earlier anecdote, I wish I could have transported my fellow delegates from Venice to the ERC symposium. I think they would have seen and appreciated the kinds of attitudes and kinds of steps one has to take to create an atmosphere for industrial competition and for economic growth.

A second anecdote, which may be well known, is nevertheless worth repeating. Recently David Packard, a man I consider to be one of our great Americans, observed to me that there are some very close parallels between success in industry and success in professional sports. He said that three factors determine these successes. One is the technical skills of individuals. Nevertheless, basic skills are essentially evenly distributed among teams, as they are among competing companies. So the other two factors make the difference in the outcome of competition. One is the individuals' zeal to win, and the other is how well they work together as a team. Few people have shown more successfully than he how those traits can be mobilized in industry, so I'm inclined to take his observation seriously. Happily, in the past few years we have seen a strong rejuvenation of that zeal to win in America, a reaction to the international pressures that we have felt on all sides.

My object in relating this story is to reinforce two points. First, we cannot play the industrial game unless we have the technical skills and the zeal to surpass our competitors, and that brings us back again to the need for a strong basic research environment, the spawning ground for ideas and talent. Second, we need better teamwork. We need to continue building cooperation and broad support for science and technology not just between the administration and the Congress, but between academia and industry too—with all accepting responsibility for making sure we nurture those technical skills and translate them into practice.

We have an exciting opportunity before us in the Engineering Research Centers. I want to put on record my strong support for what is being attempted. I hope to have opportunities over the next few years to follow their progress and celebrate their success.

DISCUSSION

A number of symposium participants from universities and industry asked questions relating to international competitiveness and the role of the ERCs. Regarding the intensification and expansion of Japan's activity in the semiconductor field, Dr. Keyworth expressed optimism about the future of American industry. Far from ignoring Japanese competition, he said, "America is rising to the competition in a very powerful and vital manner." Although capital costs and other factors will remain troublesome for the United States, technology and talent are two areas where we continue to lead. With regard to the obstructive business practices and attitudes toward R&D and competitiveness that prevail among many of our European allies, Dr. Keyworth was confident that the situation in the United States is much healthier. In particular, he noted that the extent and scope of the public debate on these issues is valuable and reassuring.

One questioner drew a comparison between the ERCs and the national laboratories. Dr. Keyworth pointed out that while the similarities are strong, the national laboratories have been concerned with meeting government requirements. He observed that the educational function of the ERCs and their location at universities gives them a different and perhaps more fundamental role.

Asked to project future funding levels and numbers of ERCs, Dr. Keyworth made several notable comments. He predicted that the current budget appropriation (for FY 1986) will be the difficult one for the ERCs to weather, but that beyond that "we are going to see monumental growth in them . . . we will be seeing units that exceed doubling for some time to come." Based on the demand for such Centers, as evidenced by the number and quality of proposals, Dr. Keyworth said he "would be very surprised if we didn't see the Engineering Research Centers become something on the order of 10 percent of the National Science Foundation [budget] in a very short period of time." He expressed his belief that the concept of a joint university-industry multidisciplinary research institute is long overdue, and that it will spread beyond the NSF to other agencies. Thus, he said, "I refuse to accept 20 [Centers] as any kind of a top."

Engineering Research and International Competitiveness

ROLAND W. SCHMITT

I believe that the main way in which engineering research and education can contribute to the international competitive position of the United States is by bridging and shortening the gap between the generation of knowledge and its application in the marketplace.

Today fundamental scientific knowledge is one of our most effective forms of foreign aid. Unfortunately, it happens to be foreign aid for our rivals—most notably the Japanese. They appreciate our research efforts so much that their industries spend two-and-a-half times as much money funding university and nonprofit research laboratories outside their nation—mainly in the United States—as they spend on such laboratories within their own country. And Japan pays us nearly a billion dollars more for patent licenses and other forms of technology import than we pay them. That favorable balance of trade in intellectual property more than doubled in the 1970s, the decade when all other balance-of-payment figures with Japan were moving in the opposite direction.

Those numbers challenge an assumption that many of us make automatically, which is that the answer to the problem of international competitiveness is to do more and more of our own research. But Japan's experience shows that it is possible to succeed in international technological competition while relying on others for fundamental knowledge and for really new ideas.

Obviously the Japanese example should not cause us to rush off and blindly imitate their methods. But it should cause us to question our accepted ideas about the relation of research to international competitive strength. That questioning could have a variety of outcomes.

One might be to conclude that we are doing the right kinds of basic research, but that we are making it too easy for our international rivals to get their hands on the results. The cure would be to put controls on the movement of our basic research results across international boundaries. Such a policy would be shortsighted. Any conceivable method of slowing down the flow of fundamental ideas to our competitors would severely damage our own creativity.

A second possible conclusion could be reached through reexamining the link between research and international competitiveness: our government might be overinvesting in basic research and underinvesting in applied research. The cure might be to shift the focus of our national research effort further in the direction of government-funded applied research and away from fundamental research. I believe this also would be shortsighted. Government must not turn from the appropriate job it does well—supporting basic research—to an inappropriate one it does poorly: trying to anticipate markets in areas where it is neither a consumer nor a producer.

ENGINEERING RESEARCH PROVIDES THE MISSING LINK

An understanding of the link between research and international competitiveness leads instead to a third conclusion. We must build on, rather than abandon, one of our greatest strengths—our fundamental research capability. But we also must ensure that it is our nation, not another, that receives most of the benefit from that strength. How can we do this? First and foremost, we must put our own fundamental advances to use more quickly than others do. We have to increase our effort in the kind of research that bridges the gap between fundamental scientific research and application. That kind of research is engineering research.

The point can be illustrated with a story. It begins in the 1880s with two German physicists, Julius Elster and Hans Geitel, who were studying electrical conduction in gases near heated solids and flames. They discovered that if they enclosed the gas and two metal electrodes in a glass bulb and heated one electrode, an electric current would flow in one direction, but not in the other. They had made one of the first electronic devices, a vacuum-tube rectifier. Yet nothing came of their discovery.

One might ascribe that failure to the fact that Elster and Geitel were pure physicists, uninterested in applications. However, at about the same time the same effect was discovered by a man no one could accuse of being uninterested in applications—Thomas Alva Edison. Edison secured a patent on one application of the effect, but it proved to be of little practical value and he dropped it.

Two decades later, in 1904, a British university engineer named Ambrose Fleming took up consulting work for the Marconi Company on the detection of radio signals. That problem inspired him to undertake some basic engineering research on the old idea of Elster and Geitel and Edison. He succeeded in using the device as a radio detector, and modern electronics was born. Furthermore, because of his ties with the Marconi Company the British were able to take advantage of the technology before anyone else did. It helped them dominate early twentieth-century radio and electronics.

Fleming was an engineer who did neither pure science nor pure engineering. He did engineering research. He was a man who knew science, but aimed to use it for a practical end. He took on engineering problems, but from the standpoint of developing generic knowledge and capabilities essential to solving those problems rather than developing products or processes. He worked in a university, but he shaped his research according to the problems brought to him by industry. He was not an intellectual pioneer like Faraday, a great experimenter like Rutherford, or a great theoretician like Dirac. But he was the right man with the right set of talents at the right time. I suggest that if England had excelled in producing and providing the right environment for many more research engineers like Fleming, just as it excelled in providing the right environment for the very few capable of reaching the heights of Faraday, Rutherford, and Dirac, the economic history of England in the twentieth century might have turned out very differently than it has.

Fleming is not an isolated example. I could equally well have chosen other engineering researchers—some operating in universities, some in industry, and some in government—such as Charles Steinmetz, W. L. R. Emmet, Benjamin Garver Lamme, Robert Watson-Watt, Frank Whittle, George Campbell, Vladimir Ipatieff, Nikola Tesla, Eugene Houdry, Warren Lewis, Gabriel Kron, Claude Shannon, Karl Bosch, and many, many more.

A Neglected Element of the Technology Development Process

The names on that list are not household words. And that is precisely the point. Engineering researchers tend to be overlooked. Our national science and technology policies are not designed with them in mind. Those policies do a good job of supporting fundamental science. Our industries do a good job of supporting engineers. And our entrepreneurs and venture capitalists do a good job of providing resources for inventors. But in the past little was done to support the work of engineering researchers in any formal way, even though they proved themselves to be enormously valuable assets in international technological and economic competition—as Steinmetz, Emmet, Kron, Lamme, and Tesla were in the electrical industries, as Campbell and Shannon were in communications, as Watson-

Watt and Whittle were in the aerospace field, and as Ipatieff, Houdry, Lewis, and Bosch were in chemistry. These people turned the practical problems of industry into exciting research challenges. They ignored disciplinary boundaries and focused instead on needs and on results; and they embedded their research in the process of innovation, rather than producing disembodied knowledge. Those are the hallmarks of productive engineering research.

The people I've named may now be history. But the role they played is more important today than ever before. That middle ground they occupied between science and engineering—the region where the leading edge of research meets the cutting edge of application—is rapidly becoming the key battleground of international economic competition. The battles over computer-integrated manufacturing, very large scale integrated circuits, communications systems, advanced engineering materials, artificial intelligence, biotechnology, supercomputers, software, and many other fields are just beginning. It is in just those fields that we will need the particular strengths of engineering researchers.

This conclusion is echoed time and again in studies by the Committee on Science, Engineering, and Public Policy (COSEPUP).* In the field of computer-integrated manufacturing, for example, the committee found U.S. efforts hampered by a pervasive lack of knowledge in such areas as geometric modeling and analysis, human-computer interfaces, and knowledge-based and expert systems. It concluded that "universities have been reluctant to grapple with the larger problems of integration," and called for universities to "educate a new breed of engineers who thoroughly understand all aspects of computer-integrated manufacturing." In the field of ceramics and composites it found that we need knowledge of structure-property relations, failure mechanisms, and design principles—knowledge that will require collaboration among mechanical engineers, chemical engineers, chemists, physicists, and materials scientists. In agriculture, maintaining American leadership will require the collaboration of agronomists and molecular geneticists. In biotechnology, the committee found that we need "a knowledge base in process engineering that combines the skills of the biologist and the chemical engineer."

Missing Elements in the Education of Engineering Researchers

We need more engineering research, and we need more engineering graduates who understand how to do engineering research. We need to put them to work in those areas where economic competitiveness is at

---

*COSEPUP is a joint committee of the National Academies of Sciences and Engineering and the Institute of Medicine.

stake; and we need to make sure that the knowledge they generate and the guidance they provide permeate the whole engineering community, not just the research community alone. We need wider and stronger bridges between the people doing engineering in industry and the people teaching engineering and doing research in universities.

In the past we have not, as a nation, paid enough attention to those bridges. The people on my earlier list did not become engineering researchers because of any role played by the government. Some did so because they could not find any other job; one did so in the course of a hitchhiking and walking trip around the world; one was a socialist escaping the persecution of a nationalist government; another was a nationalist escaping the persecution of a socialist government; one initially could not find a place on either the engineering or the scientific staffs of a major corporation, and created his own role.

What was true in those classic cases is still true today. Few engineering researchers emerge directly from the graduate schools. In some ways they resemble the religious sect known as Shakers. Like the Shakers, who were renowned for fine furniture and for the invention of the circular saw, cut nails, flat brooms, and metal pen points, engineering researchers can also claim admiration for their good works. Unfortunately, the Shakers thought natural propagation a sin, and relied on conversion alone to replenish their ranks. As a result, there are not many Shakers around today.

Engineering researchers also fail to replicate their kind. However, with them it is not a matter of morality but a matter of opportunity and inclination. It often takes years of experience at other jobs in science or in conventional engineering to turn a person into an engineering researcher. By that time he or she rarely has the opportunity or the inclination to train the next generation. Members of each generation typically are trained in a conventional engineering program, which gives them the appropriate apprenticeship for a career in engineering but not the appropriate knowledge for a career in engineering research. Or else they are trained in a science program, which gives them the appropriate knowledge for research but not the appropriate apprenticeship for making use of that research in the solution of practical problems. It is rare for a graduate student headed for a career in engineering research to be exposed in graduate school to a replica of the working conditions or professional relations that he or she will later encounter. This situation sharply contrasts with that of scientists, who are trained in the kind of laboratories in which they will later work.

THE ENGINEERING RESEARCH CENTERS: BRIDGING GAPS

As a result of these missing educational elements there is a gap between the generation of knowledge and the application of knowledge. And there is a gap between the apprenticeship of potential engineering researchers and the role they will eventually fill. The Engineering Research Centers have been designed to bridge those gaps. However, the notion of bridge-building should not be interpreted in too limited a way. The principal features of the Centers are often described as (1) industrial support, (2) interdisciplinary scope, and (3) research aimed at utility. Those descriptions are correct, but they are too narrow. They miss the essence.

Bridging Gaps Between Universities and Industry

First, the bridge established between universities and industry should carry much more than money. As one university president put it, "Don't just send us your money; send us your people who understand the critical problems. Just sending money is not enough."

Sending problems does not mean sending applied research problems. The idea is not to create Centers that are, in effect, job shops for industry. The research at the Centers should be fundamental research in the areas of engineering practice being taken on by industry—that is to say, its aim is not building robots for factories, but generating new understanding of the fundamentals of robotic vision, touch, and control; not programming expert systems for use in diagnostics or repair, but generating new understanding of knowledge representation, search and logic programming techniques, heuristics, analogies, causality, and the other fundamentals of artificial intelligence; not building biotechnology production facilities, but developing unit operations concepts for biological processes.

The goal of industry-university interaction should be the establishment of a two-way flow of information. From industry to universities should flow an understanding of the barrier problems that practice is running up against. From universities to industry should flow the knowledge and talent needed to overcome the fundamental problems. The main point is not to drive universities away from fundamental research, but to orient them toward the areas of fundamental research that are most needed by industry.

Bridging Gaps Among Engineering Disciplines

Another important feature of the Engineering Research Centers is their cross-disciplinary nature. But here again one should not take a narrow view. This is not just another interdisciplinary program; such programs more often than not simply connote a collection of specialists in different disciplines sharing office space or secretarial services. We need organizations whose shape is dictated by the problem to be solved or the type of result needed, rather than by the disciplines involved.

I am under no illusions about the difficulty that this entails. What we are really talking about is a clash of cultures: the problem-solving culture of engineering practice versus the disciplinary culture of engineering science. There will be resistance to change and suspicion of change, just as there always is whenever cultures clash.

However, in my view such an interaction of cultures does not weaken the disciplinary base; on the contrary, it strengthens it. Programs that transcend disciplines can enhance disciplinary research by revitalizing established fields and creating new ones. This is an area in which industrial research and defense research, both of which inherently transcend disciplines, have led the way. Look, for example, at the role of a one-man interdisciplinary project named Irving Langmuir and his enormous contributions to surface chemistry and plasma physics, as well as to the invention of better light bulbs and electronic tubes. Look at the contributions of interdisciplinary teams at Bell Laboratories to the solid-state sciences. And look at the revitalizing effect that highly goal-directed, interdisciplinary World War II programs, such as the ones at the MIT Radiation Laboratory, had on physics when the participants took their new-found electronics skills back to their laboratories and started applying them to nuclear magnetic resonance, high-energy physics, and radio astronomy.

These examples illustrate my point: we should not be concerned that traditional disciplinary research structures will be replaced by a new kind of interdisciplinary work done at Engineering Research Centers. Instead, we will see the emergence of new ways of doing research that will enrich strong disciplines, revitalize dormant ones, and create some new ones.

Bridging Gaps Within the Innovation Process

Finally, and most difficult of all, we must not take too narrow a view of the relation of engineering research to innovation. Instead we must seek to embed engineering research in the total process of innovation—a process that extends from identifying the market all the way through production, quality control, maintenance, and improvement of the first product into a real winner.

These parts of the innovation process cannot be separated into watertight compartments. The separation of marketing and engineering has killed many promising innovations in their early stages. Typically, the marketing people do not know enough about the future possibilities of the technology to ask the right questions of the users, and the technologists do not know enough about the users to ask the right questions of the technology. The separation of engineering and manufacturing can be just as fatal. Typically, the engineer knows too little about the possible ways the product might be manufactured to ask the right questions about the design, and the manufacturing manager knows too little about the reasons behind the design to ask the right questions about the production process.

As total-process awareness is built into the work of the Engineering Research Centers it should reflect the spirit of an experiment carried out by the late George Low, who was a prophet and pioneer of the Engineering Research Center concept. George liked to tell about a teaching program at his school, Rensselaer Polytechnic Institute (RPI), involving composite materials. To train engineers, he believed, it was not enough just to expose them to course work in the classroom and the laboratory; they also had to experience the frustration and the excitement of putting advanced technology to work. In one particular project the students conceived of a product—a glider made of new composite materials—and then immersed themselves in all the difficulties involved in "getting a product out the back door." For the final exam they were apparently required to test-fly the glider themselves! Fortunately, the glider flew. And so should the idea behind it. The Engineering Research Centers should accustom students to the idea that the engineer does research in order to do, not merely in order to know.

SUMMARY

The most effective way for us to employ our national R&D effort to improve the nation's international competitiveness is by narrowing the gap between the generation of knowledge and the use of knowledge. The place where the United States can gain additional advantage over our world competitors is the middle ground between scientific research and engineering—the domain of engineering research. In the past we have relied on chance to produce engineering researchers, and have made no concerted effort to create institutions deliberately designed to have the primary focus on engineering research. We are now designing such institutions. We should design them to create links with industry that carry not only money, but also the practical barrier problems that inspire research. They should be fashioned so as to be not merely interdisciplinary, but problem-oriented in a way that transcends disciplines. And finally, they should be fashioned so as to imbue students—and perhaps even professors—with an understanding of the true role of research within the entire process of innovation.

DISCUSSION

Two questions from the audience suggested that problems of the competitiveness of U.S. engineering are at least partly a result of shortcomings of industry. In answering, Dr. Schmitt expressed his belief that industry should not attempt to restrict publication and ownership of the results of research that it funds, and that the best way to gain commercial advantage from fundamental research is to be in a position to exploit it rapidly. He disagreed with the assertion that industry generally has trouble understanding and interacting with university researchers, or capitalizing on research with potential long-term relevance. At least in the case of large corporate laboratories this is certainly not true, he said.

To the suggestion that some ERCs might be located outside universities, he countered that universities must be the site of all Centers and that the point of the ERCs is to foster the cross-disciplinary approach in engineering research at universities. The focus on the problem rather than the discipline can be instrumental in stimulating inventiveness within the culture of the university.

Science and Engineering:
A Continuum

ERICH BLOCH

The complexity of the relations among science, engineering, and technology, and particularly the dependence of science on advances in engineering, are not well understood by scientists—or by most engineers. Science, engineering, and technology are three different spheres of activity, each with its own perspective and dynamics, yet together they should be seen as a whole, a system. Progress in each contributes to, and depends on, progress in the others.

Consider first the fundamental differences among these three areas of activity:

• There are many definitions of science, but for my present purpose I use a simple one: *Science is the process of investigating phenomena*. This process leads to a body of knowledge consisting of theory, concepts, methods, and a set of results.

• *Engineering is the process of investigating how to solve problems*. This process leads to a body of engineering knowledge consisting of concepts, methods, data bases, and, frequently, physical expressions of results such as inventions, products, and designs.

• *Technological innovation is the process that leads to more effective production and delivery of a new or significantly modified goods or service*. This process also creates a body of concepts, techniques, and data.

Some scientists believe that discoveries flowing from their work drive engineering and technology. This is true enough in many cases, but advances in engineering and technology also drive science. The "straight line" conceptual model—with progress passing from science through engineering to technology—is not only far too simple to describe the complex interactions, it is simply incorrect. Instead, we should think of a triangular model with science, engineering, and technology standing at the three corners, and vectors depicting interactions running from each of the points to the other two, always in both directions.

Differences in approach and outlook sometimes keep persons in one area from fully respecting the work of persons in the other two areas and from fully appreciating how much their own work depends on those others. This gap in understanding, in approaches and languages, sometimes appears almost as broad as the gulf between the literary and technological cultures that C. P. Snow talked about a quarter of a century ago.

Broadly speaking, scientists press for understanding, which they express as concepts, theories, and predictions. They are fascinated by the universe and its natural or social phenomena. They push forward the frontiers of their fields by finding new ways to observe, qualify, describe, and relate that part of the universe that interests them. These are clearly intellectual and creative acts.

Engineers design, invent, shape new things, make new processes, and relate concepts to solve particular problems or to uncover principles underlying a class of problems. They also strive to understand the phenomena they are dealing with, and attempt to develop the concepts and theories required to underpin their work. These are also intellectual and creative acts, no less so than in scientific research.

Furthermore, the existence of basic engineering questions and the pursuit of answers to them through research deny the common idea that engineering is only applied science. Some of the topics addressed by engineers are as fundamental to their fields as topics in basic science are to scientists. For example, research on the underlying principles of design theory, or on how to create new materials and use them in manufacturing, or on how to scale up biological processes all raise very fundamental issues.

The developers of technology, who are frequently trained engineers or scientists—although at times they are persons without much formal training—turn designs or ideas into products or services that can be used by many. They do this essentially by bringing to bear resources such as money, time, manufacturing capability, and talented people. Some of the designs, models, or ideas may have been around for a while before the developers of the technology combined them with other ideas. In addition, factors such as manufacturing costs, the potential market, and regulatory matters are taken into account more explicitly in technology development than in research.

The scientist who truly understands these differences in approach will not look down upon engineering or technological innovation, just as the research engineer or manufacturing engineer, though impatient for results, should understand that quality scientific work must follow its own dynamics.

EXAMPLES OF THE CONTINUUM

The best-known examples of the flow of ideas across and among the three areas of activity are the classic cases in which advances in scientific thought did precede and drive technological developments. The work of Townes and Schawlow in inventing the maser and laser is a good case in point. The flow in this direction is the commonly accepted model.

There are two primary ways in which engineering and technology drive science. First, the development of instruments has opened up whole new areas of investigation and given the scientist ever more powerful forms of observation and analysis. Second, many useful inventions have been developed without the benefit of scientific work, and in fact have led to the development of principles or theory—sometimes to whole new areas of science.

Many specialized instruments are crucial to advancing research—we all recognize how common lasers, computers, and other devices have become in the laboratory. And there are many more examples of technology and engineering stimulating science than might be supposed. They can be found throughout historical times right up to the present.

Some of the best-known historical examples are found in electronics, optics, and mechanics. For instance, 40 years after Volta invented the battery, Faraday finally explained how it worked. The technology of photography was worked out by artists, craftsmen, and amateurs of every sort decades before physicists and chemists understood photography's underlying principles. Perkins's work on dyes in the 1850s led to experiments in making flavorings and pharmaceuticals, which led in turn to the theories underlying the chemistry of phenols and aldehydes.

From such beginnings much of modern physics, chemistry, and biology emerged. However, we need not look that far back to see that the experiments of engineers and technology developers drive advances in scientific thought. Modern examples can be found in many areas.

The field of computer science not only arose in large part from attempts to build computers, but continues to owe a great deal to technologists and engineers—and for that matter to thousands of amateurs who develop programs and techniques as a hobby. Twice great technological developments in computers have stimulated the science of computing. The first such case occurred here and in England as part of the World War II efforts to break the German military code and to develop the atomic bomb. The second came with the revolutionary shift to very large scale integration, as miniaturization and related manufacturing processes brought with them many questions about what was going on at a smaller scale: the behavior of metals in thin layers; the surface interaction of silicon, polymers, and metals; and many more phenomena. Research in these areas has led and is leading to new scientific insights, theories, and discoveries.

The modern information era was initiated in 1948 when Claude Shannon published two papers on a general mathematical theory of communications systems. This work was based on his attempts and those of his colleagues at Bell Laboratories to track down and control noise in telephone communications channels. Shannon was an electrical engineer with a doctorate in mathematics who drew on and contributed to knowledge in both fields while solving a problem of great practical interest. Since then researchers in mathematics, computer science, information science, electrical and computer engineering, and other fields have built on his work.

Claude Shannon retired in 1972 after a long career at Bell Laboratories, having also been a visiting professor at MIT, and having won many honors, including the Medal of Honor of the Institute of Electrical and Electronics Engineers (IEEE) in 1966. I am delighted that the National Academy of Engineering recognized his work, however belatedly, by admitting him in 1984.

Among other modern examples to be found in many fields of research I will cite catalysts, which have been used in many processes for some time, with little understanding until recently of the science behind them; and pharmaceuticals, some of which were used for years before neurobiologists arrived at the modern understanding of transmitters, receptors, and blockers.

To return to my main point, then: science, engineering, and technology can properly be viewed as a continuum, with ideas, techniques, and—most important of all—people moving from one point to another in every direction.

CROSS-DISCIPLINARY WORK AND ERCs

How does this discussion of the continuum, the cross-boundary movement, relate to the Engineering Research Centers? I believe that when we look at the Centers in several years and evaluate their contributions we will find new and very significant examples of the flow of ideas and people back and forth across the disciplinary lines of science and engineering. Research in general is moving toward greater integration, more interaction. Where areas of research may converge, the Centers are designed to facilitate that convergence.

Such convergence is occurring not only among engineering disciplines, but among scientific disciplines and between fields of science and engineering:

- Biotechnology is rapidly developing as a field, but defining what it encompasses is not easy: several fields of biology, plus chemistry, chemical engineering, and physics, at least. Their interaction demands a new breed of engineers (or are they scientists?) who can synthesize ideas from, and speak the languages of, these diverse disciplines.
- Materials research is another combination of several fields of science and engineering: solid-state chemistry and physics, condensed-matter theory, metallurgy, ceramics, and polymers are some of them.
- Some areas of computer science and computer engineering are so closely allied that their boundaries are difficult to perceive. These fields are in turn contributing to—and being stimulated by—work in manufacturing systems, automation, design theory, artificial intelligence, cognitive psychology, and even bioengineering.

As the National Research Council's 1985 *Outlook on Science and Technology* points out, the fact that researchers from different disciplines are working together on common problems is not new, but the breadth of their work together is new—and so is its importance. Collaboration across traditional disciplinary boundaries, if it is to work in academia, needs strong nurturing and will require flexibility in attitudes as well as new organizational forms.

In my view collaboration should not be seen as a threat to traditional disciplines, as some people fear it to be. Work in individual fields will progress in large part on the basis of discoveries made through work in other fields, and as techniques and new instruments move from one field to another. Continuing disciplinary strength is needed as well as continuing cross-disciplinary strength. The threat I see is that university researchers do not readily understand or accept the need for cross-disciplinary work or for organizations that provide the opportunity to do such work.

Besides the involvement of scientists and engineers from many disciplines, the Centers have three other attributes that will cause their results to be widely diffused. The first is the Centers' emphasis on involving other academic institutions as affiliates. The second is their emphasis on building links with industry. The third is their emphasis on improving the teaching and practice of engineering.

With regard to involving other institutions, a college or university unable to develop and house its own ERC can become an affiliate of one. The institutions could exchange faculty members and students, and they might establish computer and video links. The resulting Center with its affiliates might be an even more productive entity, able to build on the strengths of all its components.

Affiliations can occur in many ways: institutions can submit joint proposals, as did two of the six new Centers (Maryland with Harvard, and Delaware with Rutgers); or schools sharing the geographical or topical area of a Center may join with it. I hope that as the Centers become established we will see more of this kind of cooperation and interaction.

The Centers must develop industrial partners, as experience has shown. The firms that get involved will benefit greatly from access to talented students as well as the new knowledge from research. The university researchers and students will be equally stimulated by the exchange of ideas with their industry counterparts.

As the National Academy of Engineering's 1984 report on the Centers states, each Center must assume a broad role in engineering education at all levels.* This role entails explicit efforts to codify new knowledge and to bring it to the classroom. Rebuilding the base of engineering education through modernizing teaching materials, recognizing and training teachers, and giving students the experience of participating in research is one of the most important outcomes that we can expect of the Centers.

All of us who have worked on the ERC program have very high expectations for the Centers. The Center directors and the people who will work with them face some very difficult—and interesting—challenges. Quality, not quantity, will be our guide in establishing the Centers.

Finally, those universities whose excellent proposals could not be funded because of budgetary restrictions should be urged to work with industry and with state and local governments to start Centers on their own, or to propose a Center to another government agency. The ideas in these papers can be used to improve proposals, regardless of whether they are eventually submitted to the NSF. The nation and its research enterprise will be served well by having successful and productive Engineering Research Centers, whatever the source of their funding.

DISCUSSION

Questions to Mr. Bloch focused mainly on the need for new attitudes toward and greater support for engineering. To a question regarding the relative funding for science and engineering within the NSF, Mr. Bloch replied that engineering had received one of the largest percentage increases in the Foundation's FY 1986 budget. He pointed out, however, that equality in dollars is not a good yardstick for comparison. Engineering differs from science in a number of ways, one being that it is closer to

---

Guidelines for Engineering Research Centers (1983).

industry and can therefore expect industry to contribute to its support. Viewed in this light, the NSF is really a "leveraging point" for federal dollars; both the ERCs and the Presidential Young Investigator Awards are examples of programs that leverage federal support for engineering by encouraging industry support.

Mr. Bloch agreed with an observation that engineering education has lacked the practical, apprenticeship aspect because overall support for research and teaching has been limited and engineering has not been given high priority. He noted that the ERCs, as well as cooperative and joint research endeavors among various industries and with universities, are evidence of a "change in the cultures" of government, industry, and academia with regard to engineering.

II

Genesis of the Engineering Research Centers

The Concept and Goals of the Engineering Research Centers

NAM P. SUH

INTRODUCTION

As the papers by Dr. Keyworth, Dr. Schmitt, and Mr. Bloch make clear, the concept behind the Engineering Research Centers (ERCs) is both exciting and promising. The response the Centers have received from the university and industrial communities has been overwhelming, and very gratifying. While many people made them possible, Dr. Low's role emphasizes the fact that it sometimes takes just one man with vision and imagination to influence the course of history.

To review the concept and goals of the Centers I will supplement the National Academy of Engineering report on the ERCs* and the NSF program announcement by highlighting several points.

It is appropriate to ask whether or not our mode of operation in the ERC program ought to be different now that we have gone through the initial phase. Having established six Centers, we are in a much better position to examine what we have done, and also to see whether or not the actions we have taken are consistent with the original concept.

It should be said at the outset that the final decisions in selecting the Centers were very difficult because there were so many good proposals. We used one overall criterion in arriving at our decisions: excellence. The NSF's ERC proposal review panel agreed to use excellence as the major criterion in view of the ambitious goals set for the ERC program, and in view of the enormous hope and expectations that everyone has for the ERCs.

Guidelines for Engineering Research Centers (1983).

All of us in the engineering community can be proud of the fact that the review panel experienced no political pressure in arriving at these decisions. In the final analysis, the goals of the ERC program simply reflect the goals of the National Science Foundation as established by Congress in the NSF Act of 1950. According to the act the goals of the NSF are to promote the progress of science and engineering; to ensure the nation's health, prosperity, and welfare; and to secure the national defense. In the sense of these goals the ERCs reflect our determination to strengthen engineering research and education in view of the rapidly changing international environment and the need to increase our productivity.

The goals established for the ERCs are very difficult for any institution or any nation to achieve because they require new kinds of thinking, new modes of operation, and the establishment of new kinds of relationships among our institutions. But if any organization can help the nation accomplish these goals, I believe the ERCs can, because in them we have the right people, the right institutional ingredients, and all the elements required to get the job done.

CHANGES IN THE NSF ENGINEERING DIRECTORATE

Recently we have instituted some changes at the National Science Foundation in the field of engineering. We believe these changes are necessary to meet national needs, the aspirations of the engineering community, and new requirements that may be imposed on the engineering community. Since these changes have been made to balance and complement the ERC program, a few words about the NSF's renewed commitment to excellence in engineering education and research are in order before going on to discuss the ERC program.

The NSF reorganized the engineering directorate to deal with the following issues:

- research support
- quality of engineering manpower
- facilities and equipment
- effective institutional resource utilization
- academic infrastructure for emerging and critical technologies.

We have created new programs to support research that is designed to establish a science base in fields that do not yet have such a base. We have created programs to assist universities in establishing the academic infrastructure needed to generate knowledge and trained people in many of the emerging areas in which the NSF has not had much previous activity. In addition, we have initiated ways of supporting high-risk, high-return projects even when peer review gives a mixed rating to the proposed work. These programs reinforce the traditional NSF support for engineering science and research, which will significantly affect the intellectual and technology bases of the nation in years to come.

The NSF is considering a number of other new initiatives to improve the quality of engineering manpower, basic engineering systems research, and the utilization of institutional resources such as the federal laboratories. It has a variety of programs that augment and strengthen the ERC program, and which in turn are strengthened by the ERCs. The ERC program is one of many that support university research. We are ready and we are eager to work with the university community in strengthening the research infrastructure.

RATIONALE FOR THE ERCS

One of the first questions that people asked me when I came to the NSF in the fall of 1984 was why we need the ERCs. Good answers have ·been given to that question in other papers in this volume, but I want to stress that the ERC program is a result of the realization that our engineering schools are becoming increasingly engineering-science oriented, with greater and greater emphasis on analysis of narrowly focused topics. While analysis in engineering science is an important facet of engineering, it is clear that we have neglected synthesis-oriented skills such as design, optimization of engineering systems, and system integration.

Many leaders in industry and academe complain that experimental techniques and hands-on experience are not sufficiently emphasized in our engineering schools. The way we practice engineering in industry is very different from the way we teach our students. The ERCs are needed to nurture new ideas, encourage innovation, produce better-educated people, and promote stronger interaction among our institutions, including those in industry and government.

If we do not take these tasks seriously, then 10 to 20 years from now in many of our industrial sectors we may be in a very different position vis-a-vis other countries. The ERCs are clearly a mechanism by which we can correct some of the weaknesses of our institutions today.

SELECTION FACTORS

Given these reasons for establishing the ERCs, it may be asked what specific attributes and qualifications the NSF looks for in selecting ERCs. I will just cite some of the important factors.

One important element obviously is the quality of the idea underlying the ERC proposal. Is there a new and promising idea that can strengthen engineering research and education? Is there a potential for major break-throughs, in either an intellectual or a technological sense? The overall research idea is the most important component of a Center proposal. We are looking for ideas that can produce many breakthroughs, both in academe and industry. Without such an idea at the core, a Center proposal is unlikely to succeed in being funded.

I have visited a large number of universities, and often I have been asked about the formula for success in getting ERC funding. There is no such formula. If an ERC is working on good ideas, the university will have no trouble getting industrial support; students will be challenged and interested; and universities will be able to forge the research team needed to make timely progress. Good ideas will elicit excitement.

The next element we look for in selecting ERCs is research topics. Is the problem large enough to enable a cross-disciplinary research team to work on it together and make a major contribution that cannot be made otherwise? Or does the proposal contain a collection of unrelated random topics? Is the topic relevant to meeting national needs? Are the research goals achievable?

Another element we have been looking for is the competence of the Center director and key participants. Can they achieve the stated goals of the NSF? Do they have the right mix of people? Are they capable? Do they have the needed expertise?

The fourth element in our thinking is industrial support. What is the likelihood that industry will support the type of endeavor proposed? How much support is there from industry? These questions are asked in full recognition of the fact that industrial support will vary from field to field. We have to use different measures, depending on the area of concentration.

A related factor, also important, is the type of interaction with industry. Is meaningful interaction possible? We believe that industry's participation in the research program must be substantial and real; the ERC must benefit from industrial input in all phases of its operations. Industrial participation should open up new avenues of research as well as opportunities to create new technologies. It is important that research ideas flow in two directions, from the ERCs to industry and from industry to the Centers. We believe that this ''two-way street'' quality is a vital element of an ERC.

Another element that we have been very concerned about is the educational aspect of the proposed work. Since one aim of the ERC program is to strengthen both undergraduate and graduate education, we have to ask: How are they going to involve undergraduate students? At many universities undergraduate students traditionally have not been heavily involved in research programs. If we are going to involve undergraduate students in ERCs, in what ways is it to be done, and how are they going to contribute to ERC activities? How is the experience likely to enhance their professional growth?

Still another element very much on our minds is the institutional environment. Is the proposed ERC really supported by the university? In what form and to what degree? Can the Center overcome interdepartmental barriers and actually conduct cross-disciplinary research? Are there incentive systems in place? To whom does the Center director report? Does he or she report to one department head, or to the dean? Can he or she really implement the goals of the Center, and do so through the right kind of institutional structure? We are also interested in deliverables—that is, in what a proposed Center could ultimately deliver.

To repeat, there is really no concrete formula for success in obtaining ERC funding. We are looking for creative ideas. We are hoping to be surprised by some very innovative concepts. We will even consider establishing regional Centers in areas where there are no research universities.

MEASURES OF SUCCESS

One other question that is frequently asked is: How will the NSF measure the success of ERCs? There are both short-term and long-term indicators we can employ to measure their success. Since Mr. Mayfield's paper presents short-term indicators, I will cite just a few of the long-term measures.

First of all, 20 years from now we would like to be able to see that each of the Centers has contributed in a significant way to bringing forth new ideas that have resulted in advances in U.S. engineering industries. There is an appropriate historical model. In the early 1900s a large number of universities in Europe and England, all within a 200–mile radius of Berlin, made significant contributions to physics. In fact, many of the concepts we use in engineering today came from the work of physicists in that region. One of the questions I have often asked myself is: Why was this single group of scientists able to develop so many important new ideas and principles? My answer is that they had a unique cultural environment that enabled them to interact with each other and stimulate each other's thought processes.

If they are successful and do their job right, the ERCs will help in forming an exciting cultural environment like that one—an environment that will create new intellectual frontiers and many important breakthroughs in engineering. The ERCs need to develop fundamentally new concepts and technologies comparable in scale to numerical control machining, which was first developed 35 years ago and which is having a major impact in industry today. We hope the ERCs will come up with ideas that, 20 years from now, will improve the way we live, the way we function, and the way we produce goods.

The second long-term measure of ERC success is the impact they have on our educational system. A third measure would be the impact of the ERCs on the improvement of U.S. industrial competitiveness.

NSF STRATEGY FOR STRENGTHENING ENGINEERING

Some university people are very concerned about the ERCs. They are concerned because they are afraid that ERCs will decrease support for individual research projects—that is, projects that are initiated by one investigator working with one or two students. It is my view that it would be counterproductive and a mistake to establish and fund ERCs at the expense of individual research support. The NSF has not done that, and does not intend to do it.

Funding for individual engineering research projects has increased over the past several years. In 1983 the NSF spent $82.9 million on such projects. In 1984 the amount was increased to $86.4 million (a 4.2 percent increase). Support was increased again in 1985 to $96.8 million, an increase of 12 percent. And there is $107.2 million in the FY 1986 budget for this purpose; if the Congress approves the FY 1986 request, we will realize a 10.7 percent increase over the FY 1985 level.

The NSF goal is to strengthen engineering research and engineering education in the United States. We know that we must move carefully on a broad front if we are to accomplish that objective. We cannot make the ERCs the only focus of increased funding. If we were to do that the Centers might soon act as magnets, attracting the best talent away from other institutions. That would weaken the fabric of engineering research in our engineering schools, and we must not let it happen.

The task of building strength in engineering in the United States is a very large one. To ensure that we get this strength where it is most needed we are going to have to undertake a number of new thrusts, while continuing to expand engineering research project support in the established fields. It is this type of broad-based program growth that NSF is seeking. We must have it if we are to remain a leader in engineering in the twenty-first century. It is going to take a substantial sum; I have estimated that it will cost $500 million a year.

The funding that NSF is providing for ERCs is in two parts: a minimum support element for the conduct of basic research and to maintain the infrastructure of the ERCs, and a variable support element that will depend on the performance of the ERCs, including the support they get from industry. Of course, these plans are contingent upon the availability of funds.

The NSF hopes to establish a large number of Centers. The question is, are the numbers that we have in mind enough to solve the nation's problems in engineering? My answer is that the ERCs cannot deal with all engineering problems. There are 259 engineering schools in the United States; 210 offer graduate programs, and 150 of these offer Ph.D. degrees in engineering. Even with 25 Centers we would reach only about one-sixth of the doctorate-generating institutions. Furthermore, our data show that about one-half of the 77,000 engineering bachelor's degrees awarded in 1984 were given by institutions that do not grant Ph.D.s.

What all this means is that we have a tremendous job ahead of us if we are to make a difference in the way engineering education and research are carried out in the United States. We have barely gotten started. It is apparent that we must think smartly and move ahead quickly to keep America in a leadership position in engineering. We have taken the first step. The NSF is considering a large number of other ideas that could enhance engineering education and research.

I think we can all join forces to create an exciting era for engineering and to make important contributions to the nation's industrial competitiveness.

DISCUSSION

Questions for Dr. Suh centered around NSF's plans for shaping the ERC program in the future. To a question about the possibility of funding "mini-Centers" at schools where the engineering faculty is small, he replied that NSF is open to this concept if the proposal for such a Center demonstrates that it could contribute to the ultimate goals of the ERC program. He also said that there is no policy to preclude a single university from hosting more than one ERC if subsequent proposals are strong enough on their own to win support. Dr. Suh observed that the engineering research areas represented by the first six awards should not be taken to suggest a preference for high-technology fields; mature industries such as steel-making can also benefit from engineering research. The NSF will depend on the research community for ideas to shape its strategy in this regard.

The Criteria Used in
Selecting the First Centers

ERIC A. WALKER

It is a privilege to be part of an effort aimed at strengthening engineering in the United States. The Engineering Research Centers (ERCs) are an exciting adventure, and we have great hopes for their success. Thus, I was honored and pleased to be asked to serve on the National Science Foundation (NSF) panel that evaluated the proposals for ERCs.

The role of the ERC panel is to help in the selection of the most meritorious proposals, to provide advice on ways to improve their effectiveness, and to help ensure the program's success. After I have outlined the steps taken by the ERC panel to ensure that the best Center proposals have been selected for support, I hope it will be evident that all that should have been done was done to select the most meritorious among them.

It is my good fortune to serve as cochairman of the ERC panel, along with C. Lester Hogan, former President of Fairchild Camera. Fourteen people serve on the panel; ten are from industry and four are from universities. There are a number of reasons for the heavy industry representation. One is the goal of the program itself, which is to develop new knowledge that will help U.S. industry maintain its industrial competitiveness over the long term. Another is the fact that, all together, about 300 university researchers were listed as participants in the 142 ERC proposals received by NSF. That posed potential conflict-of-interest problems in the review process because most of the university people who could function as expert peer reviewers were included in the proposals as participants.

The group brought together to serve on the ERC panel is impressive. In addition to Lester Hogan there are Willis Adcock, a Vice-President of

44

Texas Instruments; Paul Chenea, retired Vice-President for Research, General Motors; Richard Davis, Vice-President, Martin Marietta; Ernest Kuh, Professor of Electrical Engineering, University of California, Berkeley; John Hancock, Vice-President, United Telecommunications; Terry Loucks, Vice-President for Technology, Norton Company; Gene N. Norby, Chancellor, University of Colorado; Harry Paxton, Vice-President, United States Steel Company; Percy Pierre, President, Prairie View A&M University; K. Venkat, Vice-President, H. J. Heinz Company; Melvin Baron, partner and Director of Research, Weidlinger Associates; and Gordon Brown, Director, Polymer Processing, Eastman Kodak Company.

It might be wondered how such a group of people could be brought together on a panel on the same day. Pete Mayfield, who has been one of the outstandingly innovative managers at NSF for many years, accomplished this very simply by scheduling the panel's meetings for Saturdays and Sundays.

There were several steps in the review process. Before the ERC panel met, the Foundation's engineering divisions had called in 88 outside experts in the various engineering fields. These people served on topic-area panels. They reviewed all 142 of the ERC proposals submitted to NSF, and divided them into three categories—highly recommended, recommended, and not recommended. Forty of the 142 proposals came through the preliminary review with a "highly recommended" rating. The content and potential impact of the research were the principal points of focus in this review.

The ERC panel held its first meeting during the weekend of December 1, 1984. Nam Suh and Pete Mayfield each spoke during an opening session that lasted about an hour. The goals of the program as they appeared in the program announcement were emphasized, and we were briefed on what had been done in the preliminary reviews. We were given our charge, which was to identify 10 to 15 of the ERC proposals that were most deserving of site visits.

The quality of the research, the probability that the principal investigator and his or her associates would be able to accomplish the research agenda described in the proposal, and the extent to which the proposal met the goals and objectives of the program were major considerations in our review.

It was understood that the Foundation was determined to follow a National Academy of Engineering (NAE) recommendation* that the funding level for each Center be sufficient to permit the Center to make a noticeable difference in its area of research. This meant that the panel

*The NAE report *Guidelines for Engineering Research Centers* (1983) presented the NSF with recommendations regarding the establishment of an ERC program.

could not recommend 20 or 30 proposals for funding. We had to narrow the field down to a relatively small number of the very best proposals and then continue the competition through site visits to determine those that would actually be recommended for award. While each of the proposals called for a different level of funding, it was clear that only about 5 to 10 proposals could be supported with the $10 million available in FY 1985 if the ''enough to make a difference'' funding-level criterion were to be met.

We were told that we could approach the task in whatever manner we deemed best. Our first decision was to divide up and review more than 30 of the proposals that had been placed in the ''not recommended'' category during the preliminary review. We thought that this procedure would help us establish a yardstick for assessing the quality of the proposals. We also wanted to determine whether we were in agreement with the ratings made in the preliminary reviews. The panel members also scanned the proposals that were in the second, or ''recommended,'' category.

We spent several hours going over the proposals that had fallen short of the ''highly recommended'' category. Afterward there was a brief critique. This procedure proved useful, because we found that we concurred with the ratings given by the preliminary reviewers. It also allowed us to gain some experience with a number of criteria provided by Pete Mayfield. The highlights of these were:

1. The research must involve a team effort of individuals from various backgrounds, possessing different engineering or scientific skills. The research should represent an effort that can best be accomplished through cross-disciplinary research. It should not be a collection of individual research projects.

2. The Center should include a significant educational component involving both graduate and undergraduate students in research activities in the Center.

3. The Center should focus on research opportunities aimed at developing fundamental knowledge in areas critical to U.S. competitiveness in world markets.

4. There should be provision for participation by industry engineers and scientists in Center research activities. State and local agencies and government laboratories might also be participants; provision for such participation in a proposal would add to its strength.

After reviewing the ''not recommended'' proposals and discussing the reasons given by topic panels for this rating, we were satisfied that we could rely on the ratings assigned in the preliminary reviews. That made it possible for us to concentrate on the 40 proposals that had been rated as "highly recommended" in the preliminary review. We divided the 40 proposals among the panel members so that each of the "highly recommended" proposals was reviewed by at least three panel members. We then spent that Saturday afternoon and evening conducting our reviews, and convened on Sunday morning to discuss our findings.

A couple of things stood out in this review. In a number of cases the evidence of industry participation was weak. In other cases proposals were so broad that they resembled a potpourri of research without focus. But a significant number of the proposals were on target. I believe that the thing that most impressed panel members was the number of good ideas for research that appeared among the engineering research proposals. It is apparent that there is a tremendous capacity in our engineering schools for doing forefront research, and that the full capacity is not being utilized. The Engineering Research Centers will provide expanded research and educational opportunities to take advantage of that potential, and so strengthen the nation's engineering knowledge and talent bases.

Our next step was to review our own reviews of the different proposals. After discussing each proposal, we accorded it a "yes," "no," or "maybe." When we had completed that process we found that 14 of the 40 had received a "yes"; the institutions submitting these proposals were designated for site visits, which the NSF agreed to conduct.

Site teams were organized. These included at least one and usually two ERC panel members, one or two NSF staff members, and two or three consultants picked for their expertise. Some of the consultants were people who had participated in the preliminary topical reviews, and who had therefore already read the proposal.

The site visit usually included a meeting with the president and other officers of the university, who would discuss the institution's commitment to the ERC concept. There was another meeting with industry representatives. Although some time was spent visiting facilities, it was the organization of the project, the university's commitment, and certain other factors that were the primary focus of the site visits. Each site-visit team was required to write a report of its findings. There was a prescribed structure for this report that highlighted the points on which the site reviews had concentrated. In addition to an executive summary, the site-visit reports included separate sections on:

- university commitment
- management plan and capability (the longest section usually containing several subsections focusing on technical aspects of the project)
- educational components
- budget.

It is interesting to note that among the consultants who participated in the site visits were Edward Jefferson, President of DuPont, and Gordon Moore, President of Intel. The Engineering Research Centers have evidently sparked great interest in the industrial community. Our meetings with industry people revealed that there was much greater industry interest in some proposals than even the principal investigators had imagined.

After the site visits were completed the panel introduced a further step suggested by Nam Suh. Nam felt that each of the principal investigators (P.I.s) in the 14 proposals that were in the final group should have an opportunity to make a presentation to the full ERC panel. This session permitted panel members to ask questions and satisfy any unmet information needs regarding a proposal. We allowed 20 minutes for the oral presentation and reserved 10 minutes for questions. Some P.I.s commented afterward that the experience reminded them of their "orals" for the doctorate. I believe the oral presentations and the question-and-answer periods that followed were especially valuable because they gave the full panel an opportunity to learn firsthand more of the specifics of what the P.I. intended to do.

Before the oral session began, Erich Bloch and Nam Suh spoke to us again about the goals of the program. Nam Suh urged the panel to be especially sensitive to a number of factors which he called "the ingredients for success." I wrote these down. They were:

- leadership
- proper focus on problems
- bona fide industrial participation
- infrastructure, including
  - university commitment to cross-disciplinary research goals
  - internal organization
- intellectual challenge should
  - establish new intellectual frontiers
  - contribute to the knowledge base
  - provide graduate research topics
- education: should enhance opportunities for graduate and undergraduate students.

Nam said that the ERC should not be a collection of individual research topics that could be funded just as well through project grants. The panel agreed that a proposal selected for support should have the potential to achieve technological breakthroughs using a cross-disciplinary research approach. In addition, the research proposed could not be "more of the same," or simply an extension of what was already being done. It had to represent a new dimension in research in the eyes of the panel.

I believe that the principal investigators who went through this session found it to be a tough but fair exercise. We were talking to first-rate people with superb research credentials. It was a great experience for me. I believe all the panel members learned a great deal in the course of reviewing the proposals, making site visits, and sitting through the oral presentations.

The panel completed the orals at about 5 p.m. on a Saturday, and adjourned to meet again Sunday morning at 7 a.m. During the next five hours the panel members went over each of the 14 proposals. The members who had been on the site visits reviewed their findings; we studied the site-visit reports. By now each panel member knew where the strengths and weaknesses were in the proposals, and each had developed his own list of concerns about aspects of the proposals. At the Saturday meeting the principal investigators had been questioned closely about what it was that they were going to do if funds were provided. On Sunday the panel spent its time critiquing and evaluating all that it had learned about the proposals.

As we moved into the final phase, the questions most often raised were these: Would the Center, if funded, make a difference? Did it have the university and industry commitment necessary to mount a bona fide cross-disciplinary effort that would push research forward in areas of industrial interest? Was there evidence of substantial university commitment to the undertaking?

We had been asked to select the 6 best finalists and to rank the next 3. At noon on Sunday, then, the panel came to agreement on which of the 14 finalists it would recommend for NSF support. After more reviews by NSF management, including a thorough review by the National Science Board's Programs and Plans Committee, Erich Bloch made the award decisions with the approval of the National Science Board. The 6 proposals selected by NSF for funding were those that had been recommended for award by the ERC panel:

- Engineering Center for Telecommunications Research, at Columbia University
- Center for Robotic Systems in Microelectronics, at the University of California, Santa Barbara
- Biotechnology Process Engineering Center at MIT
- Center for Intelligent Manufacturing Systems, at Purdue University
- Systems Research Center at the University of Maryland in collaboration with Harvard University
- Center for Composites Manufacturing Science and Engineering, at the University of Delaware in collaboration with Rutgers University.

We were free to select proposals for award on the basis of excellence, even if that meant selecting two proposals submitted by a single institution.

The proposals we selected were, therefore, those that we believed were of the highest quality and would best achieve the goals of the program.

The ERC panel will continue as a standing body. Its role is to help the NSF select the best proposals for support, and to provide advice and suggestions on ways to strengthen the program as we go along. Its objective is to ensure that the program is a success. There is no question that the United States is being challenged as never before for technological leadership.

The Engineering Research Centers are a long-term investment. They should contribute significantly to efforts aimed at building America's engineering strength as we gear up for the competitive environment of the twenty-first century. The Centers will help improve the university infrastructure and will also strengthen the linkages between industry and universities, areas where new strength is needed if America is to continue to produce the world's best engineers.

DISCUSSION

Participants asked questions regarding the selection procedure to be used by NSF in evaluating future ERC proposals. Dr. Suh responded that the selection procedure for next year will be virtually the same as that for the first year, although NSF is seeking ways to improve the process. A new program announcement had just been issued, containing slight changes from the previous announcement.

Regarding the question of weighting systems for evaluation, Dr. Suh expressed an opinion that rating schemes are largely irrelevant, that the winning proposals stand out fairly quickly on the basis of quality of ideas. There is no set formula. Mr. Bloch confirmed that view, and added that the "believability" of a proposal is a major determinant—that is, a proposal must make clear that the interdepartmental cooperation it describes is an ongoing reality rather than an image constructed just for the purpose of the proposal.

Nurturing the Engineering Research Centers

LEWIS G. MAYFIELD

In response to the Fiscal Year 1985 program announcement regarding the Engineering Research Centers (ERCs), the Engineering Directorate of the National Science Foundation (NSF) received 142 proposals from 106 different institutions. In all, the proposals requested about $2 billion over a five-year period. Slightly more than 3,000 people were listed as participants in the proposals; 75 percent of these were from various engineering disciplines and the remainder were from scientific disciplines and the humanities.

The fact that so many institutions took the time and effort to write proposals is a strong indication of the desire on the part of engineering schools to initiate the type of research organization described in the announcement. The message must be that the format for the Centers, involving as it does both research and education on topics of importance for international competitiveness, is of great interest to engineering schools.

The total amount of funding requested by the proposals has a certain significance. The March 1985 issue of the *Journal of the American Society for Engineering Education* reports a separately budgeted engineering research expenditure in the United States of about $1.2 billion for 1983–1984. The $400 million per year requested by the proposals thus represents an increase of roughly 30 percent over current expenditures, suggesting that there is substantial unused capacity within the nation's engineering education and research enterprise. It is apparent that the engineering system has the capability and the will to perform additional research and produce more graduates without experiencing undue stress.

51

The ERC announcement was unique for the Foundation in that it called for both undergraduate and graduate students to be an integral part of the research of the Centers. For the first time universities could propose an activity to the NSF which would allow them to integrate education and research in a thoroughgoing way. This is a real step forward for engineering education and research.

NEW FEATURES OF THE 1986 ANNOUNCEMENT

The principles in the FY 1986 program announcement are unchanged from those of 1985. That is, a proposed Center should have as its focus a topic that would lead to greater effectiveness and world competitiveness of U.S. industrial companies. Proposals may be concerned with technologically strong or weak U.S. industries; there is no preference here on the part of NSF.

Several format changes have been made to facilitate both proposal preparation and review. First, a three-page executive summary is to be included. This summary will be extremely useful in the review of the proposals and will permit many more panelists to interact in a meaningful way during the review process.

Second, the section describing the proposed research program is to be limited to 25 single-spaced pages. The point is that this section needs to be well thought out by its preparers, so that reviewers can readily come to grips with the research being proposed.

The third change involves the presentation of the budget. The format for the first-year budget remains the same, but all out-year budgets must show increments above the preceding year, exclusive of equipment. This device will help everyone involved to focus on what is gained by expenditures above the preceding year.

In addition, the FY 1986 announcement encourages the formation of consortia of schools in regions where such relationships will further the educational and research objectives of the Center.

In the second round of proposals the amount and quality of industrial support will be much more important factors. In the first round there was insufficient time for proposing institutions to gain strong industrial support. I suspect that indications of industrial support will be much stronger and better substantiated in the FY 1986 proposals.

The FY 1986 ERC announcement does not include a list of potential Center topics, as the first announcement did. However, at the point when about 12 Centers have been established this "open" procedure will no longer be appropriate. When the full complement of 20 to 25 Centers has been established the subject matter they represent should encompass a broad range of research areas contributing to international competitiveness.

Therefore, the FY 1987 announcement should suggest potential topics that complement the established Centers. The National Research Council will conduct a workshop on this subject in the fall of 1985, in time to have an impact on the FY 1987 ERC announcement. At that time we will know what the proposed topics for 1986 are, and will provide that information as one input to the workshop. The important point is that we will seek guidance before suggesting a list of topics in the FY 1987 announcement.

COMMON DEFICIENCIES IN PROPOSALS

Quite a few of the proposals for 1985 had certain faults in common. Many of them were much too long. I hope that the new 25-page limit for the proposed research section will encourage brevity throughout. The reading of proposals more than 800 pages long must be considered cruel and unusual punishment for reviewers!

The FY 1985 announcement emphasized that research conducted at the Centers is to be cross-disciplinary in nature. In many instances this statement was taken so literally, and the scope of the proposed research was therefore so broad, that the research could not be adequately defined and described. Frequently the prior research of any faculty member having even a remote bearing on the focus of the proposed Center was included in the proposal. There are many potential topics for proposals which are sufficiently important and broad to meet the cross-disciplinary requirements for a Center. Setting reasonable and manageable goals and objectives would have improved many proposals considerably. Those writing proposals should keep in mind that reviewers are technical people, and that they have to feel that they understand the scope and focus of the research being proposed. Even when the scope of a proposal was suitable, many proposals failed to make an analysis of the key research issues involved in the topic.

Another major deficiency of many proposals was that they appeared to be collections of individual projects that might just as easily have been supported individually. Reviewers had to be convinced of the synergism of the projects and the people making up the proposed Center. A proposal viewed as a collection of projects simply did not make the grade. The impression that a proposal was a collection of projects was sometimes inadvertently conveyed by the inclusion of individual budgets for specific projects; in fact, on occasion these budgets were tailored to be about the size of standard NSF research grants. This approach gave a "business-as-usual" signal. I need not point out that NSF has a time-tested system for selection of individual research projects.

Still another factor that eventually influenced decisions was the leadership quality of the Center director. The perceived ability of the leadership to manage and direct the research activities of the Center was a pivotal issue during the final stages of review. The Center director has to be a versatile individual—skilled in managing people, a competent researcher, and a good leader. In addition, the Center director must be able to devote a major portion of his or her time to directing Center activities.

ERC MANAGEMENT ISSUES
"Systems Aspects"

Both the 1985 and 1986 announcements suggest that an ERC should "emphasize the systems aspects of engineering to help educate students in synthesizing, integrating, and managing engineering systems." This feature of the Centers results from the concern expressed by industrial employers that young engineers are not prepared to deal with complex engineering systems found in practice. I believe that I have a sense of the "systems aspect of engineering"; but when I talk with others it becomes clear that everyone has a somewhat different idea of the meaning of that phrase. Some think that "design" embodies the system concept; others tend to describe specific industrial problems they have encountered as "systems problems." Both notions leave out important parts of the system. While I would not deny that there is some value in having a diversity of definitions and opinions, I am made uncomfortable by the fact that the concept has not been carefully articulated.

Of equal concern to me is how best to implement education in the systems aspects of engineering. Engineering schools have an intensive program. One must ask how much more can be added while retaining the engineering science base and the humanities that we have struggled to include in engineering curricula during the past 40 years. A workshop being held under the auspices of the NRC Cross-Disciplinary Engineering Research Committee will examine this issue and prepare a report. I think that report will be studied very carefully by engineering educators and will be of considerable value.

Information Exchange

Another issue of concern to me involves methods for disseminating information from the Centers to the research community, in industry as well as universities. Is the traditional university strategy of publishing in formal journals going to be sufficient for these cross-disciplinary research centers? Should innovative techniques be developed to supplement traditional methods? At first reading this may not appear to be a very significant question; but it does have many ramifications when considered in the context of international competition. The NRC has conducted a workshop to explore this issue.

Preliminary discussions regarding this workshop have been useful. For example, I think that "information exchange" comes closer than "information dissemination" to describing the relationship that should exist with industry. After all, an important objective of the Centers is to "involve participation of engineers from industry in order to focus the research on current and projected industry needs." To accomplish this objective, universities must have a meaningful dialogue with their industrial partners. I think universities should enter into agreements with industry when the exchange of information will result in a better focus of university engineering research on current and projected industry needs. Support money may be necessary to get industry attention, but money and attention may not be sufficient if the interaction does not result in a debate leading toward more pertinent research. The question of information exchange has many facets.

Evaluation

I am frequently asked how NSF is going to evaluate the Centers. There is little question that evaluation is an important management activity. The program announcement states that three years after they are established, the Centers will be reviewed by the ERC panel to determine if each Center is meeting its proposed goals and objectives, including those for quality of research and the extent of industrial participation. This evaluation will determine whether NSF will continue to support the Center fully for the remaining two years, or provide decreased funding to terminate the Center at the end of the grant.

In preparation for the third-year evaluation the NSF Office of Cross-Disciplinary Research (OCDR) and the Center directors are preparing a list of progress indicators. These include items such as the names of graduate students at Centers, a list of Center publications, new courses attributed to the Center, and the amount and type of industrial support. This information will provide a factual base that will assist in the third-year evaluation, and that will also be useful for management purposes.

In addition to the formal evaluation, the NSF Engineering Directorate will form liaison management teams for each of the Centers. Each team will consist of a program director closely associated with the technical aspects of the Center, a program director from OCDR, a program director from the Engineering Directorate, and several outside consultants. The program director from the Engineering Directorate will be the major internal source of information for the team on the technical nature of the

Center's work. The liaison team will report to the Head of the Office of Cross-Disciplinary Research.

OUTLOOK FOR THE FUTURE

As everyone is aware, 1985 is a particularly interesting year in which to watch congressional action on appropriations. The President's FY 1986 budget, now before Congress, requests $25 million for ERC activity. With that amount NSF can maintain the Centers established this year and start a similar number next year. Of course no one can say for certain what the outcome of the budget process will be. It is easy to get caught up in the day-by-day problems and the rhetoric of the Engineering Research Center activity. It is important to note that in the long run the success of the program will be largely dependent on the quality and innovativeness of the research that is performed by the Centers, and on whether or not the students educated in the process make new and important contributions to the competitiveness of the United States.

Many people—professors, practicing engineers, and NSF staff members—have devoted large amounts of time to the preparation of proposals and their evaluation. Much remains to be accomplished, and I am confident that the good relationships developed so far between universities, industry, and government with regard to the Engineering Research Center initiative will continue.

DISCUSSION

Several members of the audience took the opportunity to ask questions relating to the proposal preparation and review processes. Not only Mr. Mayfield, but also Messrs. Bloch, Suh, Walker, and Stever responded to these inquiries. Regarding the high cost of preparing a proposal in the light of the relatively low probability of success, one questioner asked whether NSF had considered simplifying the process, perhaps by means of a pre-proposal screening stage. NSF officials responded that no change is envisioned for the near future, but stressed the importance of the proposal preparation process to the university itself for clarifying its concepts and goals governing research. NSF is trying to locate other funding sources for some proposals.

Regarding the question of what NSF might do to involve industry more meaningfully in the proposal review process, it was pointed out that conflicts of interest must be avoided—although 40 percent to 45 percent of the members of the preliminary review panel were from industry. Two options that NSF is considering are (1) to give funded ERCs a certain period of time to develop industry support; and (2) to take a long-term view of industry support, focusing on whether the essential ingredients are present in the proposal to ensure industry interest. Careful planning is necessary to ensure that a Center can continue with industry support even if NSF funding is terminated after five years.

The new 25-page limit on the research section drew some concern. Mr. Mayfield emphasized that this section should not attempt to be very detailed; instead, it should set the framework for what the proposing institution hopes to accomplish with the ERC.

Certain points in the ERC program announcement were clarified, such as the reference to "rebuilding the base of engineering education." NSF officials reiterated the need to relate engineering education to engineering practice, to codify that aspect of engineering knowledge for transmittal to students, and to help universities establish a science base in this area. The importance of this work for improving international competitiveness was clarified.

III

The Centers as a Reality—
Plans, Mechanisms,
and Interactions

Systems Research Center

JOHN S. BARAS

INTRODUCTION

The University of Maryland and Harvard University are very pleased to have been selected for an Engineering Research Center award by the National Science Foundation. On the basis of this award a Systems Research Center (SRC) will be established at the College Park campus of the University of Maryland. The focal University of Maryland organizational unit participating in the activities of the SRC will be the College of Engineering. Broad participation by several departments is planned: the Electrical, Chemical, Mechanical, and Aerospace Engineering departments within the College of Engineering; and the Computer Science and Mathematics departments, along with the Institute for Physical Science and Technology and the Center for Automation Research. The focal Harvard University organizational unit will be the Decision and Control program of the Division of Applied Sciences. In this paper I will summarize the research theme and the educational and research programs of the Systems Research Center. In addition, I will describe the planned industrial collaboration program, international program, information dissemination plans, and other aspects of the center.

The Research Theme and Its Significance

The theme of research conducted at the SRC is to promote basic research in the implications and applications of the three types of technology (VLSI, CAE, and AI)* involved in the engineering design of high-performance,

---

*VLSI = very large scale integrated circuits
CAE = computer-aided engineering
AI   = artificial intelligence

61

complex, automatic control, and communication systems. Recent advances in computer science (artificial intelligence, expert systems, symbolic computation), in microelectronics (VLSI circuits development, availability of computer-aided design tools for special-purpose designs), and in computer-aided engineering (enhanced interactive graphics, powerful work stations, distributed operating systems, and data bases) have created a unique environment for innovative research and development in the discipline known as systems engineering. For the purposes of the present paper, systems engineering is defined as the discipline that combines automatic control systems and communication and signal processing systems with certain areas of computer engineering. The major research thrust of the discipline at present is the design and implementation of high-performance electronic systems for automatic control and communication.

It is appropriate to describe some of the motivational and historical background that influenced our thinking and planning for the SRC. To begin with, the complexity of such systems has recently increased dramatically. This is manifested, for example, in tighter engineering specifications, in the need for adaptation, in requirements for multisensor integration, in the need to account for contingencies (multiple modalities), in totally digital implementations, and in the need for a mix of numerical and logical computations. Some of the challenging design problems that we plan to address in the SRC further illustrate this point:

1. How do we control systems characterized by complex, often poorly defined models? Examples from our program include chemical process control, where often it is difficult to design "correct" loops and equations.

2. How should one automate the operation of systems defined by precise, highly complex simulation models? Problems in flexible manufacturing systems in our program represent generic examples, wherein time-precedence constraints and the need for adaptive automation further complicate design.

3. How should we design systems controlled by asynchronously operating, distributed, communicating controllers? Examples from our program include the computer-aided design (CAD) of computer/communication networks, dynamic capacity allocation in communication satellites, and efficient management of mixed traffic (voice, video, data).

4. How can we develop design tools for real-time, high-performance, non-Gaussian signal processors? Examples from radar, sonar, image, and speech signal processing are found in our program.

5. How can one integrate multiple sensors for robust, digital, feedback control of nonlinear systems? Our program includes many-degrees-of-freedom robotic manipulators with vision, force, and pressure sensors, as

well as advanced aircraft flight controllers especially designed for the new generation of unstable aircraft.

The SRC will focus on the development of powerful and sophisticated software systems that will help and guide engineers in the design of automation and information-processing systems. The significance of a well-coordinated long-range research program in this critical high technology area is highlighted by the following considerations.

First, within the last year the growing role of automation in manufacturing (flexible manufacturing systems, automated factories, robotics, etc.) has attracted a great deal of publicity as the key to the health of the United States' economy and industry.

Second, an information explosion has encompassed the widespread use of computing and communication equipment (including office automation, personal computers, mobile telephone networks, distributed computing systems, sophisticated telephone networks, satellite communications, video discs, video processors, fiber optics channels, and optical storage). Among the scientific-educational community this explosion has reached across the board, from high school to university to research laboratory. More significantly, it has also been extended to the broader public.

Third, there is an increasing reliance on automatic control systems to perform precise and demanding tasks in such areas as air traffic control; advanced guidance and control systems (high-performance forward swept-wing aircraft, large space structures, and advanced space satellites); improved performance and reliability of power plants; improved control and operation of power distribution systems; sophisticated control devices for computer/communication networks; advanced electronic controllers for robot manipulators and computer vision systems; intelligent autonomous weapons and distributed sensor networks; and distributed decision systems for tactical/strategic management.

Unfortunately, currently available theories and design methodologies for such problems are not in synchrony with the currently available or planned implementation media, be it special-purpose chips or computers with specialized architectures and capabilities. More precisely, the available design theories and performance evaluation methods were developed for different (now often obsolete) implementation media such as analog circuits and sequential machines. Although for some problems—admittedly a small class—it is feasible to develop improved designs using the new hardware capabilities and existing theory, in the majority of problems there is a substantial lag between the available hardware potential and its realization in the systems being built. That gap is precisely where the Systems Research Center intends to focus.

Of course, there are examples of successful hardware solutions to some of the design problems already mentioned. By this I mean the process whereby one adds hardware components, or "boxes," in an ad hoc fashion, then tests each addition and adds more components until a satisfactory system is built. I do not believe that a serious argument can be made that this method is a superior one for exploiting the hardware potential available today. On the other hand, substantial theoretical results and knowledge exist in the form of automatic control and communication systems theories that have not been directly linked to hardware implementations. The realization that a window of opportunity exists was a major motivating force in planning for the SRC—namely, that advances in CAE, VLSI, and AI have made possible the transformation of "paper algorithms" from powerful theories into real-time electronic "smart" boxes (Baras, 1981). A careful reexamination and development of new design theories that incorporate component hardware advances and the related implementation constraints is long overdue. We can no longer separate the design of a system from the implementation problem. This is a major thrust of the SRC program.

The significance of the SRC program can also be illustrated from a financial point of view. Huge investments have been made and will continue to be made for research and development in microelectronics and computer hardware. It is important and prudent to make the comparatively small investment required for the development of design methodologies and software tools that will be used to build systems with this hardware. It is obvious that the sophistication and capabilities of the circuits and devices that we build will be limited by the power of the CAD tools that we use.

Thus, the SRC theme encompasses two fundamental components of high-technology industries: automation and communications. It is important to emphasize that high-technology industries involved in automation and communication directly influence the competitiveness and performance of more traditional industries. Consider, for example, the influence that advances in automation may have on steel mill operation and automotive design and production. This consideration was an important factor in the development of our plans for the SRC.

Educational Needs

The Systems Research Center aims at the establishment of a strong advanced research and educational program in the above areas. Given the broad knowledge and intellectual background required by the SRC research theme, we have assembled an interdisciplinary team of scientists and engineers from the two universities involved. Members of the team include electrical, mechanical, and chemical engineers as well as mathematicians, numerical analysts, computer scientists, and microelectronics and artificial intelligence experts. At its projected full operational level the SRC research program will involve some 40 faculty, 120 graduate students, and at least 120 undergraduate students. A large number of students will be influenced by the Center's educational programs. We strongly believe that there is a real need, quite critical for the nation, to educate and train engineering students in the mix of disciplines and knowledge represented by the SRC research programs. A similarly critical need exists for retraining practicing engineers, and this need will be incorporated in our plans.

THE RESEARCH PROGRAM

The research program for the SRC is an expansion and natural extension of research work already under way by members of our interdisciplinary team. The research activities listed below served as the inspiration and provided much of the motivation for the planning and implementation of the ambitious research goals of the Center. They are in a sense the seeds for interaction and further development of the key ideas behind the conception of the SRC. The SRC will provide the fertilized ground for development of the major thrusts emanating from these early works, which are:

- optimization-based design in chemical process control
- perturbation analysis and AI modeling in manufacturing systems
- symbolic computation and VLSI architectures for the design of real-time non-Gaussian detectors
- design of a VLSI DFT processor
- vision sensors and feedback in robotic manipulators.

The research program implementation selected for the SRC was influenced by three factors. First, the areas of strength of the participating faculty; second, the expected impact of SRC research; and third, a strong commitment to a problem-driven interdisciplinary program. We have as a result selected five focus-application areas to help us measure the success of the basic research program, and to help motivate it by applying the design tools to a diverse set of complex, real-world problems. These areas are described below, together with the currently planned thrusts in each.

"Intelligent" CAD of Stochastic Systems   We shall combine CAE and AI methods for the design of advanced nonlinear signal processors capable of real-time operation. One thrust is toward the development of expert systems that can "reason" mathematically and understand a variety of signal and system models. The other two thrusts address questions of

distributed computations in stochastic systems and implementation by "optimal" VLSI architectures. In particular, silicon compilation and special high-level signal manipulation languages will be studied.

Chemical Process Control   Here we shall investigate how CAE, AI, and optimization techniques can be applied to the design and control of chemical plants. Modeling and simulation questions will be analyzed and the models built, using the CAD process. In addition, we shall attempt to integrate reliability and safety considerations into the design software and work stations.

Telecommunications   There are two major thrusts here. The first centers around the development of powerful simulation and CAD systems for computer/communication networks (local-area, flow-control, and reconfigurable networks). This will involve interactive graphics, expert systems, and high-level command languages. The second thrust involves image and speech processing problems and their hardware implementation. Numerical and hardware complexity will be studied, as well as fast digital implementations.

Advanced Automation and Information Processing in Manufacturing Systems   We shall investigate applications of CAE, AI, and optimization. In particular, an integrated program will be pursued that addresses scheduling problems, adaptive resource allocation, AI systems in manufacturing, data-base integration, flexible manufacturing cells, CAD integration in manufacturing resource planning (MRP), optimization-based design, and advanced interactive simulation.

CAD of Intelligent Servomechanisms   Two major thrusts are the theory and design of an advanced prototype hand-eye machine, and the design of flight controllers for high-performance aircraft. Both involve the integration of many "smart" sensor data and the control of systems with very complicated dynamics, often requiring the use of symbolic algebra for their derivation. Implementations by special-purpose VLSI processors will be examined. In the area of robotics, the program will address primarily feedback control of a mechanical hand with many-degrees-of-freedom, based on integration of data from several sensors. In the design of flight controllers we will focus on optimization-based design for unstable aircraft.

The common thread in all these areas is their emphasis on the development of advanced CAD tools that combine the specific theory and practice of systems engineering with the three technology drivers: CAD,

VLSI, and AI. These advanced design methods provide the intellectual bond in this diversified program. The program cuts across the boundaries of a great many engineering and computer science disciplines.*

The program is interdisciplinary, problem-driven, and technique-specific. We believe that the fundamental tools, and methodologies for their design, that will be developed as a result of the SRC research program will have a very broad applicability. Furthermore, it is expected that these generic CAD tools will evolve out of strong interaction among the research activities in the five focus areas. Each area includes systems of high complexity and design problems that cannot be attacked by conventional methods. As research progresses in each area we expect to see a cross-fertilization among the various efforts toward development of CAD tools. At the University of Maryland we have already witnessed that phenomenon in design projects on chemical process control and advanced aircraft.

Still significant for the SRC's mission is the interaction between the three technology drivers (CAD, VLSI and AI) on the one hand, and, on the other, the disciplines of control and communication systems as represented in the five focus-application areas. It is anticipated that the broadly interdisciplinary program will prompt a fundamental reexamination of control and communication systems theory and methodology. Furthermore, it is expected that this latter interaction will foster a secondary level of interaction among the focus areas as hardware implementations for different applications are analyzed and compared.

Thus, the research program of the SRC will have two major components:

- in-depth investigation of the impact of VLSI, CAE, and AI
- basic research in modeling, mathematical analysis, optimization, computational and numerical methods, control systems techniques, communication system techniques, and computer engineering techniques.

The first component will address the following matters. Regarding VLSI (the implementation medium), we shall investigate algorithmic and architectural aspects of VLSI; signal processing chips; and control chips. The design methodologies to be developed must account for VLSI implementation constraints. Regarding CAE (the implementation environment), we shall investigate the effects of interactive graphics, interfaces,

---

*The disciplines include: chemical process modeling, polymers, bioreactors, chemical reactors, aerodynamics, flight controllers, robotic manipulators, vision, sensor design, signal processing, communication networks, information theory, coding, optimization, control systems design, stochastic control, detection and estimation, algorithmic complexity, algorithm architecture, VLSI array design, optimization-based CAD, numerical linear algebra, numerical mathematics, rule-based expert systems, knowledge-based expert systems, computer algebra, stochastic processes, queueing systems, manufacturing, and mechanical machining.

etc., in the design of sophisticated CAD systems. For example, in developments related to the DELIGHT Marylin system (a powerful optimization-based design system we use at Maryland), the fact that advanced graphics were to provide the output enabled the numerical analyst to develop an interactive procedure that could handle multi-objective optimization. In addition, this environment permits the engineer to study a design problem in his own language, without being overburdened with complicated computer procedures. Regarding AI, we shall investigate the effects of symbolic computation and knowledge-based systems on design.

The second component is needed because sophisticated new theories and methodologies are required in order to extract the maximum benefit possible from advances in microelectronics, CAE, and AI. As Roland Schmitt (1984) describes the situation: "In the technology of controls, . . . fundamental theoretical advances are needed to catch up with the speed and power of microelectronics."

The impact of VLSI technology on signal processing and automatic control systems is emerging as very influential. However, for success in this direction very advanced CAD tools must be developed and popularized. The rapid developments we have seen in VLSI chip design and production were made possible by the development and rapid dissemination of precisely such advanced CAD tools. The SRC program aims at producing similarly sophisticated CAD tools in the general area of control and communication systems engineering design.

An important factor in future systems engineering theories and design techniques will be the development of expert systems for CAD (Stefik and de Kleer, 1983). In applying expert systems to design tasks the idea is to pit knowledge against complexity, using expert knowledge to whittle complexity down to a manageable scale. It is anticipated that expert systems will eventually be applied in many design areas; but their use in digital system design, particularly in CAD, will be a major advance. The planned SRC program will develop a broad research activity in this area.

AI and symbolic computation promise to revolutionize design. There are very sound reasons for this prediction. First, the cost of generating special-purpose Fortran-based codes is fast getting out of hand. Massive investments in design tools can become either a brake on innovative designs or an argument against further development. AI symbolic computation transfers mathematical models of the physics of the system being designed from the code side (applications code) to the data side of the system, where they can be used, manipulated, shared, modified, and even created by the system as easily as numerical data elements. This transfer is essential for the attainment of cheap, easily reconfigurable design tools. Second, AI and symbolic computation prevent the designer's entrainment in specific design procedures and processes provided by custom-coded

Fortran programs, and thus allow for a very flexible approach to the design. Symbolic manipulation has immediate and powerful applications in CAD. For example, the amount of nuisance programming required to develop and maintain large design packages can be reduced to a practically negligible part of the overall code. Furthermore, increasing the level of abstraction at which data and code are specified reduces significantly the complexity of code transportability. Finally, symbolic manipulation permits entire mathematical models—logic as well as its numerical parameters—to be treated as data capable of being manipulated, examined, and modified, as well as being executed like a Fortran subroutine. Further advantages offered by AI include natural language processing, automatic deduction, cognitive models, and learning and inference. An excellent example for an application of AI and symbolic computation in aerospace design is given in Elias (1983).

Systems engineers today are called upon to solve complex control and communication design problems for systems often described by huge simulation models. The traditional approach has been to reduce the complexity to a small number of mathematical equations and eventually apply rather simple elements of available theories. Clearly we can do much better than that if we utilize the full power of techniques from CAD and AI. Furthermore, the speed provided by VLSI arrays promises to support the often real-time processing need of advanced control and communication systems. For systems of the complexity seen today it is often difficult to write and manipulate the governing equations correctly. Think, for example, of the task facing a chemical engineer who is trying to describe a complex industrial chemical process, starting from simple, elemental chemical reaction equations. His final goal is to design a process controller. Or consider the aerospace engineer who is developing a mathematical model for a large, complex, multi-body, flexible structure in space. Again, his final goal is to design a controller. Both have to manipulate a large number of equations (often more than 100) of different types (algebraic, differential, partial differential, Boolean, etc.). Symbolic manipulation and rather elementary AI techniques (such as search heuristics, "sup-inf" decision procedures, etc.) can readily reduce these tasks to routine and permit the engineer to concentrate on the design issues. More generally, there is clearly an established need for utilizing AI methodologies in CAD. In the design of flexible manufacturing systems, for example, one encounters coordination problems that can benefit enormously from the use of automated reasoning programs. The latter can supervise the lower-level numerical CAD programs. To ask such a systems engineer to solve the complex design problems of today without such a combined arsenal of tools is similar to asking a VLSI chip designer to design the chip without the expert CAD tools now available.

In this discussion I have tried to convey the basic ideas behind the research program of the SRC. However, research is not the only purpose of the Engineering Research Centers. Our plans for the SRC educational program are equally important.

THE EDUCATIONAL PLAN

The basic theme of the SRC educational programs is that the Center supports and enhances educational programs and is a source of new courses and material. Furthermore, the SRC and the two universities involved are committed to the principle of lifelong education (see Bruce et al., 1982).

The SRC will establish within the first two years a modern environment for rapid information dissemination via a local computer/communication network, appropriately connected to the University of Maryland university-wide network and other industrial and government networks. Adequate work stations and advanced terminals will be provided to support sophisticated computer-assisted instruction tools. Software libraries, case studies, and design examples will be maintained and updated. We plan to utilize, in a timely manner, powerful educational tools such as video discs, video tapes, personal computers, and work stations. The recently initiated university-wide drive for such an environment will accelerate and support this effort. Similar efforts are under way at Harvard University, and the electronic linking of these two educational networks will establish a superb educational environment.

The Harvard University faculty group will participate in the development of course material. We plan to maintain these materials in computer files (in a "troff" standard format) and to exchange them, together with other course materials developed at Maryland, through computer mail. This will permit rapid revision and reproduction of lecture notes and other materials.

Each research project at the SRC will generate a research seminar on related background and research topics. The seminars will be flexible, and will attempt to produce lecture notes on the research performed. Successful projects and seminars will then endeavor to produce polished versions of these lecture notes for publication and wide distribution.

Seminars will involve graduate as well as undergraduate students. In fact, we plan to utilize the software systems developed by the SRC as educational tools for students. This will serve two important purposes: timely codification of new knowledge and research results, and continuity in training and education for the students participating in the Center.

In the current planning, all courses affiliated with the Center will be initially of the seminar or independent study type, and closely linked to ongoing research projects. Undergraduate students will participate in every research project.

The SRC will maintain strong and active visiting programs for scientists from both academic and industrial as well as governmental research laboratories. Outside scientists will be utilized extensively as instructors.

The Center will organize an advanced-level retraining program for practicing engineers from industry, government, and other institutions. In addition, cooperative programs will be established with industrial affiliates that will provide summer employment to students at government or industrial research laboratories.

Dissemination of research results from the SRC will be implemented via technical research reports, a quarterly technical magazine, software systems, and lecture notes on specific subjects. In addition, the Center will hold yearly working seminars and colloquia on focused research areas (one per year), with wide participation from outside scientists and students. The Center will also utilize short, intensive courses (7 to 10 days) and the University of Maryland's Instructional Television (ITV) System in order to popularize and disseminate its research findings.

The educational plan of the Center is designed to blend naturally with the academic offerings of the participating units. Periodic reviews, performed in collaboration with participating academic departments, will ensure that information and knowledge are transferred to the regular academic curricula in a timely manner.

INDUSTRIAL COLLABORATION PLAN

There is growing acceptance of the fact that technological and industrial innovation are central to the economic well-being of the United States. Universities are one of the major performers of the fundamental research that underlies technological innovation. Industry puts this research to work, and also identifies problems requiring new knowledge. The flow of people and information between the campus and industry is thus an important element in both scientific and technological advancement. A broad collaboration with industry in the research areas of the SRC is expected. Industry-university partnership has become a national objective, and this new environment is particularly helpful to the Center's program. The significance placed on a healthy and strong industrial collaborative program is manifested in the establishment, within the SRC, of an industrial liaison office for managing this program.

Corporations with strong research and development (R&D) activities in areas related to the SRC research program will be invited to join in a group known as Systems Research Affiliates (SRA). Its purpose will be to provide for continuous strong scientific and educational interaction and to support the Center's activities. There will be three grades of membership in this group, depending on the kind and level of involvement with SRC activities: sustaining affiliate, sponsoring affiliate, and affiliate.

The Harvard University team will participate in this activity by assuming responsibility for the initiation of contacts with high-technology firms in the Boston area. It is anticipated that such contacts could lead to involvement of these firms in the activities of the Center, both at Harvard and at Maryland.

Industrial participation will occur in many modes: joint research projects (both with and without private funding); exchange of scientific personnel for limited periods of time; sharing of advanced equipment; joint development of a software library and "software club"; use of private industry laboratories and test beds for SRC projects; specialized education for practicing engineers; cooperative employment programs for SRC students; work-study programs; fellowship programs; and industrial funding of equipment, students, faculty, and workshops at the SRC. Our strategy for developing a strong industrial collaboration program is based on the building of strong technical ties with industrial research engineers. The collaboration between the two universities, and the unique characteristics of their respective regions, offer distinct advantages to the SRC. It is a somewhat innovative feature of the planning for the SRC that a synergistic research, development, and education effort will be undertaken in a critical high-technology area by the three most concerned communities: university, industry, and government.

Industry will be appropriately represented in the administrative, management, and research structure of the SRC. There will be industry representatives in the administrative and research councils of the SRC, professional resident fellows from the sustaining affiliates, and visiting scientists from industry.

It is worth reiterating the importance of university-industry collaboration in the programs of the SRC. First, we believe that a key to innovation is a close coupling between the researchers and developers of technology and its users. This coupling must be in place during the entire innovation process. Second, certain of the technology drivers of the SRC, such as VLSI and AI, have been vigorously pursued by industrial research labs because of their enormous potential commercial value. Third, lack of skilled manpower is particularly acute in the mix of disciplines underlying the SRC. Strong industrial research participation in the proposed SRC will enhance considerably the probability that the Center will succeed in its mission.

CONCLUSION: A FORMULA FOR SUCCESS

The Systems Research Center links two major universities that possess broad intellectual, engineering, and scientific expertise. It also links two major and complementary high-tech centers. Two exceptionally strong and complementary groups of systems scientists and engineers will collaborate in an ambitious program that is expected to interact with and impact all other Engineering Research Centers established to date. In addition to the $16 million in NSF support for the first five years, the SRC has strong commitments from both universities and from the State of Maryland. The University of Maryland has committed 12 new faculty positions, 10,000 square feet of new space, $1 million in operating funds, $1 million in dedicated equipment, and another $1.6 million in shared equipment. Harvard University has committed two new faculty positions, 1,550 square feet of new space, and computer networking. The Maryland Department of Economic and Community Development will assist with the establishment of the SRC and, in particular, its industrial collaboration plan. The SRC will collaborate on research, education, and industrial programs with the University of Maryland Engineering Research Center (established by the State to provide an engineering extension service, equipment grants, and an incubator facility), with the University of Maryland Institute for Advanced Computer Studies, and with the National Security Agency (NSA) Supercomputing Research Center at the Maryland Science and Technology Park.

I will end with a brief recapitulation of our goals. The future design of electronic control and communication systems must be performed by sophisticated computer-aided methods, all the way from problem definition to blueprint specifications for the electronic circuit or the software implementing the design. This vertical integration should be accomplished in the next decade, and is indeed the driving goal in the research program of the SRC. With the utilization of expert systems and CAD tools across different application areas, substantial cross-fertilization occurs as the experience and expertise pass from area to area in an unending loop, each time improving the power and efficiency of the design methodology. This type of interaction is not possible without the utilization of these technologies. We believe that the interdisciplinary program of the SRC will spark a fundamental reexamination of control and communication systems design as we develop the design tools for the electronic "smart" boxes of the future.

I emphasize again that the proposed research program includes a substantial component of fundamental research in mathematical modeling, analytical studies in optimization and dynamics, sophisticated methods from statistics and probability, and advanced computational methods and numerical algorithms. These are essential if we are going to demand that the designs produced offer a substantial improvement in performance over conventional ones. Indeed, we hope to develop CAD tools that will be able to produce special-purpose software and hardware implementations utilizing very advanced, albeit expensive, technology. Without a sophisticated analytical/computational foundation, the advisability of such designs is questionable. To put it simply, engineers are going to need very advanced analytical/computational tools in order to squeeze out of the final implementation (be it hardware or software) every possible performance improvement. CAD is clearly the economical way to go—the alternative being a sequence of untested trial-and-error experimental designs. Essentially, it affords extensive testing and evaluation at a low cost.

The harmonious marriage of powerful analytical/computational methodologies with the three technology drivers described in the SRC research program summary is bound to produce what we would like to call the ultimate systems engineering technology. Finally, the SRC is dedicated to the education and training of a new generation of control and communication systems design engineers.

REFERENCES

Baras, J. S. 1981. Approximate solution of nonlinear filtering problems by direct implementation of the Zakai equation. Pp. 309–311 in Proceedings of the 20th IEEE Conference on Decision and Control. New York.

Bruce, J. D., W. M. Siebert, L. D. Smullin, and R. M. Fano. 1982. Lifelong Cooperative Education. Report of the Centennial Study Committee, Department of Electrical Engineering and Computer Science, MIT. Cambridge, Mass.

Elias, A. L. 1983. Computer-aided engineering: the AI connection. Astronautics & Aeronautics 21 (July/August):48–54.

Schmitt, R. W. 1984. National R&D policy: an industrial perspective. Science 224 (June):1206–1209.

Stefik, M. J., and J. de Kleer. 1983. Prospects for expert systems in CAD. Computer Design 22 No. 4 (April):65–76.

Center for Intelligent Manufacturing Systems

KING-SUN FU, DAVID C. ANDERSON,
MOSHE M. BARASH, and JAMES J. SOLBERG

SUMMARY

The focus of Purdue University's new Engineering Research Center is on intelligent manufacturing systems—a phrase intended to describe the next generation of automated design/manufacturing systems. That generation will emerge in the 1990s as fully integrated, flexible, self-adaptive, computer-controlled systems covering the full range of factory operations from product concept through delivery. The Center is organized to support the long range cross-disciplinary effort in research and education needed to bring this technology into wide use in American industry.

Twenty to 30 professors will conduct the research, working in project teams with as many as 120 graduate students and, through a novel arrangement, another 80 undergraduates. National Science Foundation (NSF) funding for the program is approximately $1.6 million dollars for the first year; the amount will increase year by year, totaling as much as $17 million dollars over the first five years.

The new Center for Intelligent Manufacturing Systems builds upon Purdue's success with the Computer Integrated Design, Manufacturing, and Automation Center (CIDMAC), which broke new ground in organizing cross-disciplinary, joint university-industry research. The new Center complements the activity of CIDMAC to achieve broader technical coverage and wider industrial impact. It also addresses the problems of educational innovation that must accompany a new style of engineering.

INTRODUCTION

This Engineering Research Center represents a serious effort on the part of Purdue University to address the needs of the American manufacturing industry in meeting competitive challenges. It is motivated by an awareness of both the importance of a healthy manufacturing base to society at large and the essential role of engineering education and research in meeting industrial needs.

It is now widely recognized that the American manufacturing industry is facing competitive pressures that literally threaten its survival. Furthermore, it is clear that our engineering education system must undertake significant revisions to meet the new challenges. Recent studies, such as that of the National Research Council's Committee on the Education and Utilization of the Engineer, have clearly identified and characterized these needs.

The overall problem is systemic in nature. Small-scale, narrowly focused programs dealing with one or another of the many aspects of the problem cannot adequately confront the global issues involved. Four overriding requirements stand out:

1. There is a need for new mechanisms to provide closer *linkage* between industry and the academic community, in order to close the implementation gap and ensure the smooth flow of information in both directions.

2. Greater emphasis on the *integration* of engineering knowledge (in both research and education) is needed to deal with the increasingly pervasive systems nature of problems.

3. The methods and structure of engineering *education* must be renovated in order to provide the kind of engineer that industry will need in the future.

4. A *large-scale cross-disciplinary effort* will be required if the program is to have sufficient impact to benefit a wide spectrum of manufacturing practices at a national level.

Taken together, these requirements dictate a significant departure from the traditional behavior of most universities. Even with sincere, dedicated efforts by highly competent people there are countless ways to fail. The viability of any effort of this kind depends critically upon a well-selected focus, a wisely constructed organization, and a carefully planned system of program management. Figure 1 depicts the way in which the Center is embedded within the Purdue University organization. The relationships and functions of contributing entities will be described in later sections of this paper.

FOCUS OF THE CENTER

The new Center will focus on intelligent manufacturing systems. The phrase "manufacturing system" is sometimes used in a narrow sense by

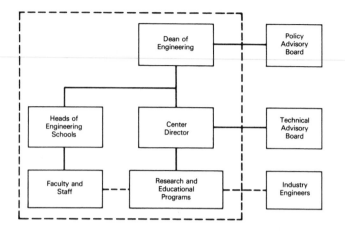

FIGURE 1   Place of the Center in the Purdue University Organization.

machine tool suppliers to refer to integrated production machinery. The meaning here is broader. It encompasses all those activities associated with making products, including design, planning, processing, management, inventory control, and every other aspect of the product realization process. Although the kinds of products considered will not be limited, our primary emphasis will be on mechanical and electromechanical products that are made in discrete units. The continuous-process industries (e.g., paper and steel) have unique problems that we do not intend to address, although the Center will address some generic issues that would apply equally well to both discrete and continuous production.

Background

To gain a better understanding of what the word "intelligent" means in the context of manufacturing systems it is useful to trace the historical development of manufacturing. From the earliest times through the first stages of the Industrial Revolution the dominant feature of manufacturing was human labor. Later, when mechanical automation was first coming into common use, the center of attention was on fixed automation for high-volume production; the conveyor belt is an example. Beginning in the 1950s numerically controlled machines and electronic process controls began to take over some of the low-level supervisory burden from people. Starting in the 1970s and continuing today, the emphasis has been on flexibility (i.e., the ability to adjust to changing requirements) and integration of computer software. It could be said that we have passed though three generations of manufacturing technology and are now in the midst of the fourth generation (see Figure 2). Looking ahead to the next 20

1st generation (pre-20th century)—manual labor
2d generation (early 20th century)—fixed automation
3d generation (1950s)—numerical control
4th generation (1970s)—flexible automation
5th generation (1990s)—intelligent systems

FIGURE 2   Stages of evolution of manufacturing technology.

years, what novel features are likely to characterize the fifth generation of manufacturing?

The historical evolution of manufacturing over the past century reveals two major long-term trends. First, there has been a shift in emphasis away from labor and machinery concerns toward more abstract information concerns. Today the single most significant feature of manufacturing is the complexity of the information involved. Thus, a major new problem faced by all manufacturing businesses is how to engineer properly the information structures that feed and control all of the product-realization processes. We may expect fifth-generation manufacturing systems to utilize major advances in the technology of information engineering.

The second major trend is an increasing awareness of the significance of systems phenomena. Whereas early manufacturing sought to segment, or decompose, the many steps of product realization in order to simplify the management of the overall enterprise, it is now recognized that many problems can be attributed to inadequate attention to the interfaces between these steps. The gap between design and manufacturing is a notorious one; but there are many other examples as well. Fifth-generation manufacturing systems will involve a far greater integration of technologies to achieve true system optimization.

The term we use to describe this next major advance in production technology—this fifth generation—is "intelligent manufacturing systems." The words are intended to carry the connotation of higher-level computer control, flexibility, and integration. All sectors of manufacturing will be impacted by these changes, some sooner than others. The Center will conduct research, education, and technology transfer activities to facilitate expeditious progress toward the goal of greater control, flexibility, and integration.

Research Focus

What would an intelligent manufacturing system entail? Where should a university-based research program focus its attention? In order to develop a comprehensive picture to guide consideration of these questions, consider first the manufacturing functions themselves. These can be conveniently grouped into three categories: design, processing, and planning and control. Together, these three labels cover all the functions that transform concepts, requirements, raw materials, and resources into products. Although they are carried out in various ways and with varying degrees of success by different industries, it is clear that the three categories are logically distinguishable and that all are absolutely essential to the realization of products.

In addition to participating in these functions directly, engineers in industry have concerned themselves with supplying equipment to improve these processes. We will call this kind of improvement process "automation engineering." It is external to the manufacturing function itself, in the sense that it is more concerned with modifying the processes than with making products. Historically, the attention given to automation has been hardware-intensive, and generally localized to individual components of the larger manufacturing system. Even the most modern flexible manufacturing systems can be fairly characterized in this way.

Our focus on intelligent manufacturing systems requires a new emphasis on the information aspects of the system. Information engineering is similar to automation engineering in its objective to improve the direct manufacturing functions; but it is distinctly different in its emphasis on logic, procedures, organization, and software instead of equipment. Because it is impossible to deal with these latter aspects in isolation, information engineering also requires a completely integrated approach to the entire manufacturing system, including all of the pieces mentioned above. Figure 3 illustrates the overall concerns of our research program, emphasizing the integrated nature of the elements (this is not the only way to structure the Center's approach, of course).

Detailed technical work will be carried out in all of the areas. However, it should be understood that the wholeness of the program is more important than any one part; it is this feature that distinguishes the Center for Intelligent Manufacturing Systems from a mere collection of projects. A typical project within the Center will involve the faculty of two or more disciplines dealing jointly with research issues that cut across traditional boundaries.

THE RESEARCH PROGRAM

The traditional reliance upon individual investigators or small teams of investigators works quite well for well-defined single-discipline projects. There is no need to criticize that approach; any other that we might plan

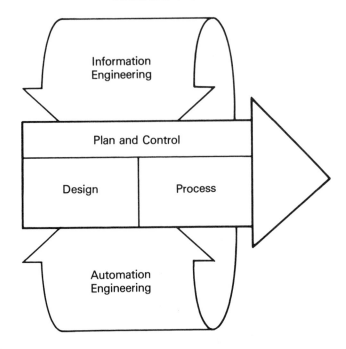

FIGURE 3   Elements and interactions in the intelligent manufacturing system.

would needlessly undermine it. However, a cross-disciplinary Center offers opportunities for research that are not otherwise possible. One of the principal reasons for conducting research in a Center organization such as ours is to coordinate activity within the group so as to achieve more than separate projects alone would permit. That is, the synergism itself has value. Moreover, the dialogue about research priorities that automatically accompanies the organization of such a Center serves as a self-regulating, unobtrusive force to ensure relevance of the separate components. This is particularly true when industrial representatives with knowledge and vision can enter the dialogue, as has been and will continue to be the case in planning for our Center.

Of course, a cross-disciplinary Center introduces a need for program management of a kind not ordinarily faced in conducting academic research. Aside from the mechanical details of organization and operations, it is important that clear guidelines, or a philosophy, be established for selection of the specific research topics to be investigated. Obviously, the topics should have intrinsic merit, a strong cross-disciplinary flavor, and a clear relationship to the focus of the Center. Beyond these points two additional guidelines should be followed.

First, it is important to be very selective about the projects undertaken. Although this Center will represent an undertaking of considerable magnitude, the total effort can be no more than a small fraction of the national research effort in manufacturing. We cannot hope to do more than contribute to the overall effort. Furthermore, the needs are so great in so many areas that it is pointless to compete with other universities or research institutions who are already excelling in certain areas. Generally, we want to pursue those projects that are in line with our central mission and that we are best equipped to undertake in terms of background, talent, facilities, and circumstances. We will concentrate our resources and attention on exploiting our present strengths.

A second guiding principle in selecting research projects is to maximize leverage. This Center will identify problems that offer the possibility of a very large payoff for a reasonable amount of work. These problems tend to coincide with the most critical needs in industry—the bottlenecks to productivity—but are not necessarily short-term or easily overcome obstacles. Generally, universities are best suited to working on the frontiers of knowledge, where the commercial incentives are not yet clear enough to encourage private initiatives.

Although most engineering research is carried out with the conviction that it will eventually prove beneficial, the manufacturing arena imposes special requirements to make explicit the costs, benefits, and timing of research. The fact that the benefits may be long-term does not diminish the responsibility of engineering research for consciously addressing these issues, because the ultimate penetration of any innovation in manufacturing depends as much on economic factors and timing as it does on technical success. Therefore, all of the Center's research projects will be evaluated in terms of their potential economic benefits as well as their purely technical aspects.

It would be counterproductive to detail in advance a rigid, permanent structure for all of the research to be undertaken. Creating an environment that encourages innovation requires a flexible organization that is clearly and openly receptive to new ideas. Furthermore, the target we are trying to hit is moving. It is possible to provide a sample listing of likely projects to indicate the general flavor of the work the Center will support. Of course, it will be the responsibility of the individual investigators to form teams and organize their own proposed efforts, which will then undergo review before funding. Some likely research areas are:

- computer-integrated system for product design/analysis
- data-base system for CAD/CAM integration
- automatic intelligent process planning and production preparation
- "virtual manufacturing" software for intelligent manufacturing

- process technology integration
- intelligent error compensation in precision machining
- intelligent control for automated production and assembly
- integrated microsensors for intelligent control
- advanced intelligent assembly and inspection.

These are only examples, of course, and even such inherently cross-disciplinary project clusters as these must address the long-range objective of total system integration.

THE EDUCATIONAL PROGRAM

In education as in research it is essential to have a guiding philosophy. It is neither necessary nor desirable to build an educational program from scratch. The existing engineering programs have evolved over years and have developed strengths that it would be foolish to ignore. On the other hand, it would be equally foolish to pretend that our engineering educational system is adequate to meet the new demands that are emerging from concerns about national competitiveness. In concert with the research program and with the more directed technology transfer program to be described later, our Center will explore fresh approaches to providing the human resources that will be needed in the future. It will do this through a combination of traditional and nontraditional educational programs.

It should be noted that many of the education delivery mechanisms that are just now coming into vogue nationally have been in place at Purdue for some time. We have had a widely praised engineering co-op program enrolling more that 1,300 students annually, and involving approximately 650 corporations. An undergraduate program in interdisciplinary engineering, involving about 125 students a year, provides an opportunity for custom-tailored curricula to meet special needs in such areas as biomedical engineering, transportation, and energy. We have long had an extensive continuing engineering education program. Our on-campus conference facilities are among the largest of such facilities at any university and are fully utilized. Television instruction has been fully developed for both on-campus and statewide off-campus learning.

Notwithstanding this strong background of programs in place, we recognize that new initiatives are needed to focus and structure an educational program that specifically addresses manufacturing requirements. Because the subject is evolving rapidly, the program must stress fundamentals rather than fads. It should foster the spirit of innovation, and prepare the student to live comfortably in a world of constant technical change. We should strive to make the program attractive to the very best students, because the regimen is likely to prove more demanding than a traditional program.

It is easy to identify narrowness of disciplinary perspective as one of the major educational problems in engineering today. It is also easy to structure a cross-disciplinary program that permits students to pick and choose from a wide selection of courses. What is difficult is to build a cross-disciplinary program that avoids superficiality.

Thus a new curriculum plan is needed, drawing on the expertise of several engineering disciplines. No existing engineering program can handle this challenge alone. Furthermore, the programs that are most closely related to the need are already hard-pressed to satisfy the demands for teaching and research in their traditional areas of expertise. New initiatives and new resources are needed.

Aside from the curricular changes that are already under study at Purdue, and which will be aided by the new Center, we have devised a plan that squarely addresses the dilemma of how to involve undergraduates directly in research—something they normally have little opportunity to do. It is a difficult problem, because undergraduates have little time to devote to research while meeting graduation requirements, and they usually lack sufficient in-depth knowledge to have much to offer in advanced work. However, we have devised a strategy to permit the most capable students to become genuine members of a research team.

The Center will offer to qualified students the opportunity to become Summer Undergraduate Research Interns (SURIs). During one summer, probably between the junior and senior years, a SURI will join an ongoing research project team (along with faculty and graduate students who have been working throughout the year). He or she will be paid a salary of up to $2,500, and will be expected to make a positive contribution to the research. To qualify the student must have taken certain courses (beyond the usual required courses) and have earned good grades. These qualifications are to ensure both adequate preparation and seriousness of intent. This is a position to be earned, not a financial aid program. We expect as many as 80 SURI positions to be awarded, and have included the cost of student wages in the budget.

The SURI program is highly experimental. If the plan can be made to work, it promises benefits to both the students and the research projects. Moreover, it may serve to encourage the best students to pursue graduate studies.

At the graduate level, Purdue has developed and is currently in the process of implementing a Masters-level core program in manufacturing systems of engineering. This program, which was developed jointly by the Schools of Engineering under a grant from the Westinghouse Foundation, emphasizes the integration of technologies of design, manufacturing, and automation. The program will be administered by the Schools of Engineering, with guidance by a policy board consisting of the heads of the Mechanical, Electrical, and Industrial Engineering departments. An advisory committee of corporate representatives will be formed. This committee will participate in major policy discussions and review program plans. Academic administration will be handled by individual schools and will be consistent with graduate school policy.

In the category of continuing education, Purdue has recognized the need for new delivery mechanisms to address the problem of technical obsolescence of mid-career engineers. The president of the university has spoken often of the need for fresh approaches to this increasingly important challenge, and has declared it to be a high priority item for the entire university. The Center will participate in the full range of new offerings that are being developed toward this end. Beyond these initiatives the Center will develop its own program, to be specifically targeted at engineers in the manufacturing industries most affected by our research.

INDUSTRIAL PARTICIPATION

Fulfillment of the mission of this Center will require the direct participation of industry. Financial contributions are important, of course, as they are needed to support Center operations. The financial commitment also ensures the continued active attention of important people in the companies. But it would be a mistake to think of the relationship as merely a financial transaction, in which a company is buying something of value from Purdue. Rather, the arrangement should be considered a joint venture, in which all parties contribute and share. CIDMAC has proven that such a concept can work.

Mechanisms

The new Center for Intelligent Manufacturing Systems will offer two levels of industrial participation. The existing CIDMAC program will become a part of the ERC, and will be expanded to include more member companies. This form of participation requires a significant commitment because it involves maintaining a close working relationship, including a full-time site representative. The site representative is a technically oriented employee of the company who resides at Purdue and engages in the day-to-day activities of the Center. In addition, each member company will have representation on the Policy Advisory Committee and on the Technical Advisory Committee. It is understood that the individuals on these committees represent not just their own companies' interests, but those of American manufacturing industry as a whole. There is a reasonable limit to the number of companies that could participate in this manner— probably in the range of 12 to 15.

The other form of participation in the Center is as an "affiliate." Affiliates receive a newsletter and reports, and may attend an annual research forum. Generally speaking, they will be observers of the activities of the Center rather than direct participants in the research. The obligation of an affiliate to the Center involves no more than the payment of an annual fee.

Technology Transfer

Compared to other industrialized nations, the United States has not provided very well for routine technology transfer between universities and manufacturing companies. This fact alone is one of the major reasons for lagging productivity gains. Any comprehensive strategy that is intended to improve U.S. manufacturing productivity must address this issue.

The on-site industrial representatives of our principal industrial partners, as well as the affiliate program just described, will serve as effective conduits for the exchange of information and technology arising out of the Center's work. Other, more commonplace means of information/technology transfer, such as workshops and conferences, will be organized from time to time as appropriate. The usual practices of publication and presentation will also be carried out. The philosophy that must guide all of the research done within the Center is that it be available to all U.S.-based companies. Although our industrial participants may enjoy special advantages by virtue of their direct involvement in the work as it occurs, the Center will not restrict dissemination of results to just those few companies. Indeed, it is part of the mission of the Center to actively disseminate its results so as to improve the competitiveness of American manufacturing industry.

Center for Robotic Systems in Microelectronics

SUSAN HACKWOOD

INTRODUCTION

The Center for Robotic Systems in Microelectronics, located at the University of California at Santa Barbara (UCSB), brings together two technologies of vital importance to U.S. industry: robotic systems engineering and microelectronics manufacturing. By working in this cross-disciplinary area the center will generate advances in applied as well as fundamental research.

The main goals of the Center are to create new technology in flexible automation for semiconductor device fabrication and to educate a new generation of engineers skilled in the implementation of robotic systems.

The program being implemented at UCSB involves faculty and students from four different engineering departments, and has a unique method of interacting with industry. The educational program now under way will produce graduate and undergraduate students who will be familiar with the needs of industry, and who will be capable of designing and building automation systems.

UCSB, located about 100 miles north of Los Angeles and 250 miles south of Silicon Valley, is geographically well situated to become a major research focus for California's high-technology industries. A high level of university commitment, along with National Science Foundation (NSF) funding, will ensure the success of the Center during the start-up period. However, the eventual goal of the Center is to become self-sufficient through funding from industrial sources.

ROBOTICS AND MICROELECTRONICS

We define "robot" as a computer-controlled machine which is self-reprogrammable via sensory inputs. Mechanical arms are examples of robots; but within this broad definition semiconductor processing equipment can also be regarded as a robot if the equipment is endowed with sensory-based control.

The pursuit of automation in microelectronics can have two distinct goals. The first is process investigation. Here robotics is used to control a fabrication sequence automatically and reproducibly to obtain optimum results. The second goal is increasing yield, while maintaining quality and reliability. As very large scale integrated circuit (VLSI) features continue to shrink (1-micron features will be standard in the future) and complexity increases, production costs will be the critical factor for competitiveness in microelectronics. Thus, of these two aspects of automation in electronics—process investigation and increasing yield—the Center will emphasize the latter, although without neglecting the first. This emphasis was chosen because productivity is economically the most critical aspect, and because it is not currently being researched.

Accordingly, we have chosen research areas that will result in reduced costs in semiconductor device fabrication. To accomplish this, robotic systems that allow more accurate alignment, reduction of particles and defects, and better control of complex processes are being developed. Three general research areas have been selected for focus: (1) robotic systems for material transfer, (2) robotic systems for process control, and (3) robotic systems for assembly and packaging.

Many universities have realized the extreme importance of robotics research and education for the survival of our economy. Unique to the UCSB Center is the emphasis on systems. We define a robotic system as "a collection of interacting robots and peripherals that together achieve a definite purpose." Recently, it has become apparent that the bottleneck in robotics is not so much in the science as it is in the implementation.

The United States may still lead in the fundamentals, but it is lagging behind Japan in system design and application of robotics in industry, as Figure 1 makes clear. The Center, while advancing the basic knowledge of robotics in mechanics, in control, and in sensors, will stress the integration of robotic systems into real-life environments. The result, as suggested by Figure 2, should be an accelerated reduction of manufacturing costs and an improved U.S. competitive position.

MANAGEMENT AND RESEARCH METHODS

The Center is led by a three-person team. The type of university-industry cooperation envisioned for this Center requires a range of leadership tasks

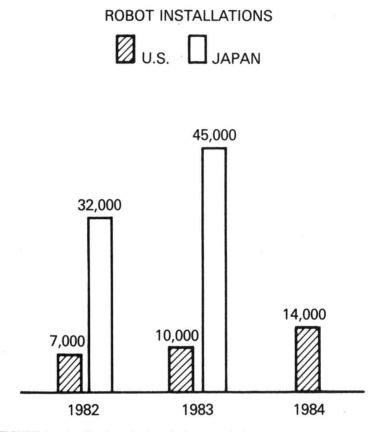

FIGURE 1   Application of robots in Japanese industry compared to application in U.S. industry.

that are best carried out by a team. In this way it is possible to maintain the executive effectiveness of the leadership without sacrificing its technical expertise. (Such a sacrifice is inevitably made whenever the executive function resides in one person alone, as is often seen in the case of university leaders who, in order to administer, have irreversibly lost contact with research.)

The Center also proposes a new method of engineering research. The usual procedure is to go from the general to the specific, to do first research and then development. Typically, research is done freely in academia, and out of the results produced industry picks those worthy of development. Robotic systems research cannot be done this way. The procedure used by the Center is to go from the specific to the general, doing applications first and gaining fundamental knowledge later (see Figure 3).

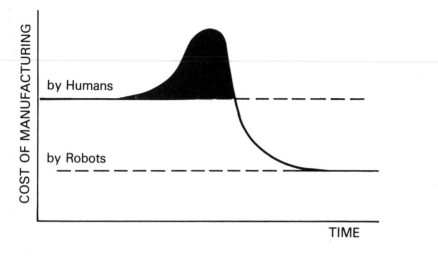

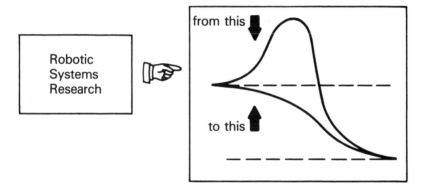

FIGURE 2   Effect of robotic systems research on speed of application of robotics to U.S. manufacturing and on cost reduction.

Research focused on real applications does not, however, deny the importance of pertinent basic underlying principles.

Research is carried out by a multidisciplinary team of 16 professors from four departments (Electrical and Computer Engineering, Mechanical Engineering, Chemical Engineering, and Computer Science). Projects are under way in the basic disciplines of robotics (mechanics, control, and machine perception), as well as on applied research in flexible automation of microelectronics manufacturing.

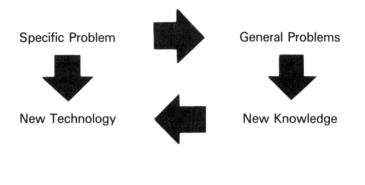

(Short-Term)   (Long-Term)

FIGURE 3   The Center's approach to robotic systems research, involving the creation of new knowledge by generalization from specific tasks.

INDUSTRIAL INTERACTION

"Systems House" Approach

A strong interaction with industry is the key to the success of the Center. The Center operates as a "systems house." This systems house is the missing link between the robot builder and the robot user. The concept involves several key features. First, an automation project is selected jointly by a private company and the Center faculty. The project is chosen for its importance to the company, relevance to the advancement of robotic systems engineering, level of difficulty, time scale for execution, and amount of industrial commitment. Project design and execution occur at the Center. Industrial participation includes the company's assigning engineers to work with the Center. Implementation takes place on location in industry.

In the systems house mode of operation the Center can become financially self-sufficient in the following way. The company purchases the equipment necessary for the system to be implemented. The equipment is then loaned to the Center. Equipment is loaned rather than donated because the end product of a research effort is the successful transfer of the same equipment, in the form of a complete system, back to the company. In return for the completed system the company is asked to fund students and faculty for the next project in proportion to the complexity of the system. The total cost to industry is substantially less than the cost of commercial systems houses, and the university benefits since the Center receives student/faculty support and is assured of always having the most up-to-date equipment available for research, at no cost.

Current Industrial Participation

At present the Center is supported by approximately 15 companies. These include: (1) large semiconductor and/or equipment manufacturers (Intel, Rockwell, Varian, and Bell Communications Research); (2) local industries (Renco, Circon, Santa Barbara Research Corporation, and Delco); and (3) robot manufacturers (GMF, Intelledex, Automatix, Microbot, and Digital Automation Control).

One project that has already been initiated, in collaboration with Bell Communications Research, is the development of a standard way to handle long-wavelength semiconductor lasers. Semiconductor lasers are expensive ($200 to $400) because the production yield is so low. This project is aimed at designing robots capable of inspecting, testing, and handling these fragile devices. In particular, a four-axis, modular micromanipulating robot with vision capabilities is being constructed. This is also an example of self-support achieved via the systems house method of operation. Bell Communications Research will purchase the system upon completion.

FACILITIES

The university has leased an Engineering Centers building at the edge of campus. The Center for Robotic Systems in Microelectronics will occupy 14,000 square feet of this space. A major new acquisition of the Center is a 1,400–square-foot class 100 clean-room, which has just been purchased by the university and will be installed by mid-June. These facilities greatly enhance the Center's chances of success. A further 4,000 square feet of space in the newly constructed engineering building on campus will also be allocated to the Center, for undergraduate teaching.

The university has allocated approximately $1 million for initial equipment purchases, and has also allocated several faculty positions to the Center.

EDUCATION

The Center is stimulating the teaching of new courses in subjects relevant to robotic systems in microelectronics. Both undergraduate and graduate courses are offered. The research program will involve approximately 10 percent of all engineering graduate students at UCSB. Multidisciplinary engineering methods are assimilated by example, through participation in the university-industry joint projects. The Center will stress undergraduate education. It has already established a one-year, senior-level complete curriculum for robotic systems specialization. The Center provides funds to pay undergraduates as technicians in the implementation stages of projects. It will also make extensive use of videotaped instruction to illustrate robotic systems implementations on the factory floor. In addition, the Center is open to members of industry to continue their education, and invites the participation of other schools in the same geographical area.

Center for Composites Manufacturing Science and Engineering

R. BYRON PIPES

The College of Engineering of the University of Delaware, in association with Rutgers University and with funding by the National Science Foundation, will develop a Center for Composites Manufacturing Science and Engineering. This Center is intended to provide cross-disciplinary engineering research and training to support national needs in the commercial aircraft and aerospace industries, the ground transportation industry, the electronics industry, and other consumer products industries.

The initial impetus for a national emphasis on composite materials came from the need for new materials to meet the extreme and exacting requirements of the aerospace programs of the 1960s. The demands on materials in these applications were so diverse and severe that no single existing material could satisfy the requirements. The development of new stiff, strong, and lightweight materials systems, consisting of high-performance fibers unified by advanced binders, played a key role in the success of the space program as well as in the development of new military systems. Today, while such materials continue to be important in space and military applications, they are also being required to play much broader technological and economic roles with regard to national needs in the commercial sector.

OVERVIEW: THE CENTER'S GOALS AND CAPABILITIES

The new Center intends to provide a cross-disciplinary approach to the conduct of engineering research and to the development of engineering graduates at the bachelor, master, and doctoral levels. As a partnership among university, government, and industry, the primary goal of the new Center will be to accelerate technological advancement through discipline synergism, scholarship, sustained basic research, graduate and under-graduate education, unique facilities, faculty excellence, technical exchange, and documentation of the evolving technology. The primary roles of the university center will be in the development of new knowledge and the transmittal or transferal of knowledge and technology. Since universities are not self-sufficient in the evolution of technology, the active participation of industry is imperative.

The University of Delaware has been a pioneer in the development of the center concept. The Center for Composite Materials, founded in 1974, was the precursor of the new Engineering Research Center (ERC). In addition, a Center for Catalytic Science and Technology was founded in 1978 and has developed a strong national program in catalysis research, supported by the National Science Foundation and the petrochemical industry.

The facilities developed under ERC sponsorship are maintained for the joint use of all faculty and students involved in Center projects. A professional staff is provided for maintenance of research facilities and tools. The Center provides services not typically accessible to students and most faculty, such as graphics services, a research professional staff, technician services, enhanced clerical service, and access to unique facilities.

OVERVIEW: THE RESEARCH PROGRAM

The research program of the Center for Composites Manufacturing Science and Engineering will focus on fundamental engineering research problems that represent the primary barriers to the growth of this important new high-technology industry. Five primary research programs make up the Center: (1) Manufacturing and Processing Science; (2) Mechanics and Design Science; (3) Computation, Software, and Information Transfer; (4) Materials Design; and (5) Materials Durability. The interaction between design and manufacturing science in composite materials requires the careful integration of the first two programs, while the remaining three programs will form the cross-disciplinary foundation of the Center. The affiliate program of Rutgers University will allow for extension of the research to encompass ceramic matrix composites, in addition to polymeric and metallic systems.

Manufacturing and Processing Science

Contemporary manufacturing processes all share one characteristic: greater control of material microstructure brings higher manufacturing costs. Yet it is through the careful control of material microstructure that the greatest improvements in material properties can be obtained. Significant opportunities for composite materials lie with the development of automated manufacturing methods whereby control of microstructure can be achieved. Thus, the primary objective of the Manufacturing and Processing Science research program is to develop the fundamental engineering and science basis to support the development of manufacturing methods for composite materials.

Particular emphasis will be given to development of active control of material microstructure and properties. Three primary areas of focus will be manufacturing process studies, quality assurance and nondestructive evaluation, and fundamental process variables. Net shape forming processes to be studied include: robotic fiber placement, laminate sheet forming, injection and compression molding, textile forms, powder processing, pultrusion, and reaction injection molding. Quality assurance and nondestructive evaluation studies will focus on the simultaneous monitoring of in situ material properties and defects utilizing sensing techniques that include optical, piezoelectric, and radiation sources and sensors. A robotic work station will be developed for computer-aided interrogation of complex geometric forms. Tomographic, holographic, and advanced ultrasonic techniques will be utilized in this program. Fundamental process-variable studies will examine the rheological properties and processes of composite materials. In addition, the development of material microstructure will be examined through such studies as the measurement of crystallite dimensions in semicrystalline polymers and cross-link density in thermosetting polymers.

Mechanics and Design Science

The Mechanics and Design Science program will develop mechanics models for several emerging composite material forms of interest, and will integrate the models into a computer-aided design methodology. The material forms will include: textile structural composites, ceramic matrix composites, flexible (elastomeric) composites, and hybrid composites. For each form, constitutive relations will be derived and the failure process modeled. Computer-aided design science research involves the integration of not only materials models, but also processing and manufacturing science models. In this way the computer-aided design research will permit the simultaneous design of material microstructure and external geometries.

Computation, Software, and Information Transfer

The availability of high-speed, large-capacity computers has changed many of the traditional approaches in engineering education and practice. The heuristic methods of the past have given way to numerical simulation, which permits the solution of field equations that describe the complex, coupled phenomena involved in manufacturing, processing, and design of composite materials. To advance this process, the Computation, Software, and Information Transfer program will develop computational models for the prediction of material behavior, and will provide for transfer of technology to industry by means of computers. Accordingly, this program emphasizes research on computational analysis, materials modeling, advanced computer graphics, computer-aided design, and materials data base.

Materials Design

The aim of the Materials Design program is to generate concepts and methodologies that link materials processing to performance. This end will be accomplished through coordinated research efforts directed at relating process-induced variations in a hierarchy of structures to material behavior. The structural-hierarchy approach offers the potential of connecting molecular structure to macroscopic behavior through the coupling of the behavior of key structural elements associated with differing scales of interaction. At the macrocomposite scale the arrangement of reinforcing elements is considered. The focus at the microsystems scale will take into account inhomogeneities in the internal structure of the reinforcing agents and matrix, as well as the possibility of a perturbed interphase region near the surface of the reinforcing element. The development of explicit molecular theories to describe the properties of ordered regions of crystallitic materials will be considered in the molecular systems research effort.

Materials Durability

The Materials Durability program is directed at the rational design of composite materials to prevent premature failure. Primary thrusts of the research are, first, to define microscopic failure detail experimentally and thus produce microscopic failure models; and second, to develop quantitative computer models relating microscopic detail to macroscopic failure. The two primary material forms considered are continuous and discontinuous fibers embedded in homogeneous matrices. Phenomena investigated will include those associated with the actions of the environment, mechanical stress, electrical stress, and foreign body contact. These include rupture, creep, wear, fatigue, and dimensional instability. Characterization of the initial and degraded states of the material, using the advanced tools of electron microscopy, infrared spectroscopy, nuclear magnetic resonance, dynamic mechanical spectroscopy, local chemical measurement (ESCA,* Auger, or x-ray fluorescence), and x-ray or light scattering, will permit proper model development.

Ceramics Research at Rutgers University

The Affiliate University program involves faculty and students of the Ceramics Department of Rutgers University in studies to enhance toughness of ceramic materials through fiber reinforcement and to develop methods for injection molding of ceramic preforms from powder starting materials. The benefits to the two universities will be substantial in that the University of Delaware expertise in composite materials will be combined with that of Rutgers University in ceramics; thus, the overall program will be expanded to add the important class of ceramics to those of polymers and metals.

ACADEMIC PROGRAM

Embedded in the educational program of the College of Engineering, the new Center will involve undergraduate students, beginning in the sophomore year, in participation as undergraduate research assistants. In 1985 a total of 30 undergraduate students will be employed to assist graduate students, faculty, and/or research professionals for 10 hours per week each. In the summer after the sophomore year students are to be engaged full time at the Center, while in the summer after the junior year they are to be placed in an industrial or federal laboratory to gain practical research experience. Senior students normally elect to conduct an independent research effort under faculty guidance. Considering the special opportunities it represents, this program for undergraduates is directed toward the academically accomplished student, and provides a vehicle for recruitment to graduate programs.

The focal point for involvement of students in the Center for Composites Manufacturing Science and Engineering will be as graduate research assistants. In 1985 a total of 33 graduate students throughout the College of Engineering will be supported by Center funds, and by 1990 a total of 50 graduate research assistants will be active in Center programs. As degree candidates in the curriculum departments of Chemical Engineering,

---

*Electron spectroscopy for chemical analysis.

Civil Engineering, Electrical Engineering, Materials Science and Metallurgy, and Mechanical and Aerospace Engineering, the students will receive specialization through cross-disciplinary research projects and through specialized course work. Research projects will culminate in theses and dissertations, as well as research progress reports.

Primary objectives of the educational program will be the development of Ph.D. degree-holders to carry out industrial and federal research and to set up similar academic programs at other universities, along with M.S.- and B.S.-level graduates to carry out engineering practice in the emerging composites industry. The interdisciplinary awareness of these graduates should be a key factor in their success in each of these endeavors. Finally, the educational program will provide for the continuing education of both engineering practitioners and young entrants to the field from other professions through short courses and specially prepared text materials.

INDUSTRIAL INTERACTION

Industrial participation in the Center for Composites Manufacturing Science and Engineering will have a pervasive influence upon the program. It will take many forms: direct financial support for facility development, financial support of consortia programs, financial support of individual research projects, participation in advisory boards, and exchange of personnel through industrial internships.

Financial support through a joint University/Industry Research Program known as "Application of Composite Materials to Industrial Products" will provide approximately $1 million per year. Industrial funds will also be provided for the purchase of facilities for a Composites Manufacturing Science Laboratory (CMSL); the initial investment will be approximately $1 million, to be provided during the first two program years. Ten blue-chip companies are participating in this way. Exchange of personnel will be extensive. The residence of industrial personnel within the Center for periods of 6 to 18 months to conduct joint, open research with Center personnel will provide an important mechanism for interaction with the Center. In addition, a Visiting Scholar Program will provide for the placement of university faculty or research professionals in industrial or federal laboratories. It is anticipated that, in all, 30 to 40 industrial organizations will interact with the Center in various ways through the University/Industry Research Program.

Three mechanisms are provided for industrial review of Center programs. They are an Industrial Advisory Board, a Manufacturing Science Advisory Board, and a Science Advisory Board. Membership in the Industrial Advisory Board will be open to industrial organizations who join the University/Industry Research Program described above. This board will be comprised of seven subcommittees: research, technology transfer, computer software, student honors, patent policy, long-range planning, and facilities. Vehicles for the transfer of technology to the industrial sector include the production of a composites design encyclopedia, annual workshops, an annual research symposium, computer software, site visits, and industrial internships.

FUTURE DEVELOPMENT PLANS

To support the development of the new Center, the University of Delaware will establish three new tenure-track faculty positions in the College of Engineering. Support for the new faculty positions will be borne by the Center during the life of the program; the university will take financial responsibility upon program completion.

The Center for Composites Manufacturing Science and Engineering will expend approximately $4 million from 1985 to 1990 in the development of facilities to support the research program. The renovation of more than 6,000 square feet in Newark Hall will provide for new laboratories: a Nondestructive Evaluation and Quality Assurance Laboratory, a VAX 11–785 Computing Facility, a Publications Production Laboratory, and the first phase of the Composites Manufacturing Science Laboratory (CMSL). Construction of approximately 13,000 square feet of new space will provide for an office and laboratory structure, as well as for completion of the CMSL. Approximately $1.5 million will be spent in support of equipment for the new laboratories.

Engineering Center for Telecommunications Research

MISCHA SCHWARTZ

SUMMARY

The Engineering Center for Telecommunications Research was established May 1, 1985 at Columbia University by a major grant from the National Science Foundation (NSF).

The focus of the Center's research efforts will be on integrated telecommunication networks of the future. Two major thrusts are planned. One is on developing new systems and concepts for these networks, which will handle, in an integrated fashion, data, voice, video, and other communications traffic. Technological advances in very large scale integrated (VLSI) circuits, microelectronics, and electrooptical devices will be required to achieve the degree of integration we are proposing. These needed advances provide the second, and concurrent, thrust.

To explore the network aspects of integration, we are implementing a highly flexible network test bed called MAGNET, which is capable of supporting data, facsimile, voice, and video communications. At the same time we are developing work stations designed to access a network such as MAGNET, thus providing an interactive multimedia environment with real-time voice and video as well as data and graphics. Our microelectronics and electrooptical devices group has begun development of some novel electrooptical devices. New laser-beam microfabrication techniques will be used to build these devices. We plan to explore and implement new VLSI and multimicroprocessor architectures in order to meet the processing demands posed by real-time voice and video traffic within the work stations and network switches.

We expect to involve 200 undergraduate and 200 graduate students in course work, project work, and research activities associated with the Center. New courses and curricula of a multidisciplinary nature will be developed, based on Center activities. Industry will be closely involved: appointments will be made to an industrial advisory board; new adjunct positions will be created; an industrial visitors program will be established; and short courses for industry will be developed.

INTRODUCTION

The U.S. telecommunications industry is one of the largest in terms of gross product; it is also among the world leaders in the development and use of high technology. The field has been expanding explosively world-wide, and it is now at a critical juncture in its evolution because of two recent developments with far-reaching significance. First, during the past decade the marriage of communications and computer technology, together with the accelerated pace of breakthroughs in microelectronics and lightwave technology, have produced a proliferation of new devices, systems, and services, ushering in what is often termed the "information age." Second, the competitive environment in the U.S. has been changed fundamentally by deregulation, by the AT&T divestiture, and by impressive advances in high technology on the international scene—Japan being the outstanding example.

What have been the consequences of these developments? Until very recently the U.S. was the undisputed leader in telecommunications research, and a few large industrial organizations dominated the scene. Under those circumstances the output of a small number of industrial and university research laboratories was sufficient to maintain that dominance. Today the situation is changing rapidly and dramatically. All computer manufacturers have entered the telecommunications field very actively, adding to the already heavy involvement of the traditional carriers and communications manufacturers. The field is now wide open to competition, and non-U.S. manufacturers—from Canada and Japan particularly—are making important inroads into U.S. markets. Members of the European Common Market—the French and British in particular—have declared modern telecommunications to be a top-priority high-technology field. They are ahead of the U.S. in a number of areas of telecommunications services, and are also looking for ways to penetrate U.S. markets. German, Swedish, and Italian manufacturers are also actively involved. The new competitive atmosphere has spawned a wide variety of companies, many of them very small, that are actively engaged in the development of telecommunications products and services. Finally, in the face of the increasing complexity of the field as well as the new opportunities it offers, the large users of telecommunications services (the prime example being the financial community) have been driven more and more to develop their own systems and expertise. The result of these several trends is that an acute need has developed for the expansion of telecommunications research in the open atmosphere of the university, for the transfer of the results of this research to all industrial organizations—large and small— and for the training of a much larger group of specialists to keep pace with the proliferating needs of industry.

How should the university respond to this challenge? An impressive array of disciplines is involved in the art of telecommunications. For example, knowledge of optics, acoustics, microelectronics, the psychophysics of perception, and the mathematics of signal processing is needed for applications such as voice and image processing, recognition, and understanding. Queueing theory, combinatorial mathematics, economics, and law all contribute to the conception of new systems and services. It is clear that future advances in the state of the art will require an integrated approach demanding the combined efforts of systems engineers, theoreticians, and specialists in solid-state devices, lightwave technology, VLSI circuit design, computer hardware and software, and other fields yet to be identified. Significant investment in laboratory facilities and support personnel is also required. (The communications industry worldwide has long recognized the need for this type of investment, as is illustrated by the massive efforts in VLSI technology at AT&T Bell Laboratories and at NTT and NEC of Japan.)

In view of these characteristics of the field, it seems self-evident that an effective university organization for telecommunications research must be multidisciplinary in nature, with a strong experimental component and a close working relationship with industry. Particularly because of the experimental aspect of the work, such an organization requires a fairly high level of funding; and in order to insulate it from the exigencies of the industrial arena, it should be largely (although not necessarily exclusively) funded by government. We believe that the Engineering Center for Telecommunications Research, established at Columbia through a major grant of the National Science Foundation, meets these requirements.

THE RESEARCH PROGRAM

Overall Research Focus

The key concept in future telecommunications systems is that of providing integrated services for a variety of interconnected users. Future telecommunication networks will carry data, voice, graphics, facsimile, video, and other types of traffic in such a way that they are "transparent" to the user. New systems and new concepts will be needed to make these networks possible. Integration within the network must go hand in hand with integration at the user interface. Future user terminals are expected to have built into them a real-time voice and video capability, in addition to the ability to handle data and graphics. Basic studies of this multimedia environment are required. Our Center research activities will focus on developing new concepts in integration within the network and at the user interface.

Achievement of the degree of integration we are proposing will require technological advances on a number of fronts. The high-speed data-rate requirements set by integrating various forms of traffic dictate the use of optical transmission. Novel electrooptical devices will be required, integrating optical and electronic processing on the same chip. New laser-beam microfabrication techniques will be used to build these devices. The processing demands posed by real-time voice and video traffic within the user terminals and network switches require orders-of-magnitude increases in processing power over existing systems. New VLSI and multimicroprocessor architectures will be required to meet this challenge.

It is apparent that the research activities outlined above are multidisciplinary in nature. The research to be carried out ranges from the basic physics of materials and processes to the mathematics of systems analysis. Although the goals of the research are specific, focusing on integrated telecommunication networks of the future, the implications are broad and include the exploration of new directions in man/machine communication and in auditory and visual perception, as well as new means of organizing information services.

In order to carry out these research activities most effectively the Center is organized into four major activity areas:

- systems and new concepts
- VLSI circuits and architectures for telecommunications
- microelectronics and electrooptical devices
- analytical studies.

Eighteen faculty members of the Columbia University School of Engineering and Applied Science are participating in Center activities. Departments involved are Electrical Engineering, Computer Science, Industrial Engineering and Operations Research, and Applied Physics. Students, full-time research staff, and industrial visitors will also participate in the four activity areas.

Investigators in the systems and new concepts area will be exploring new concepts in integrated network architectures and integrated work stations (terminals). Work in both image processing and speech compression will be carried out as part of this integrated services effort. Researchers in the VLSI circuits and architectures area will develop integrated circuits for telecommunications as well as new, automated techniques for reducing the functional specifications of a telecommunication system component to a circuit layout on a chip. They will work closely with the systems people on implementation of some of the new concepts and systems developed, as well as on VLSI architectures for image and voice processing for real-time transmission over the integrated networks.

It is noted above that electrooptical devices and lightwave technology will play a key role in research activities focused on integrated networks of the future. The researchers in the microelectronics and electrooptical devices area will be involved in a number of activities important to this aspect of telecommunications. These include studies in lightwave technology and laser fabrication technology, the development of microelectronic devices for high-speed signal processing, and the design of new optical devices. The analytical studies group will carry out studies fundamental to an understanding of telecommunication network performance and design. These studies are expected to provide feedback on and ideas for new concepts in integrated networks. A network simulation facility will be developed to provide additional support for these activities.

MAGNET: An Example of Current Research Activity

In beginning our studies of integrated networks we are implementing a highly flexible network test bed called MAGNET. MAGNET is a local area network of our own design capable of supporting integrated services such as data, facsimile, graphics, voice, and video communications. Through proper software design it will also emulate, at higher levels, integrated networks of various types. Once completed, it can be used to study integration of services on local area networks, as well as to provide a test bed for trying out new system concepts as they are developed.

The initial implementation is based on coaxial cable technology. Concurrently with the development of MAGNET, the electrooptics group is developing the optical components that will enable the network to be switched to fiber optic technology. The fiber optic implementation, consisting of two fiber optic rings, each operating at 100 megabits/sec transmission capacity, will enable wide-bandwidth video signals to be transmitted over the network, in addition to voice and data. There are plans to have 12 nodes on the network. These include a powerful minicomputer and several intelligent microcomputer work stations.

To provide a truly integrated network—that is, one that integrates data, voice, and video from the point of origin through the network to the destination—integrated work stations must be available. Commercial work stations available to us in our laboratory do not have this capability. Work is therefore proceeding on novel speech compression and image processing algorithms to enable work stations to handle real-time voice and video. Integration of real-time video services, in particular, is a formidable task. These studies will enable us to explore, with much better understanding, the requirements for integrated and multimedia work stations of the future. They should lead to the novel VLSI and multimicroprocessor concepts required to fully implement the integrated services environment that the future will bring.

EDUCATIONAL/INDUSTRIAL PROGRAMS

Along with these research activities focusing on new concepts for integrated telecommunication networks and the technology required to support them, the Center will also be pursuing a variety of related educational activities. Student project work at both the undergraduate and graduate level will be expanded considerably through the facilities of the Center. Full-time research staff, funded through the Center, will work jointly with faculty on a multidisciplinary basis to guide students in the conduct of projects and formal doctoral research studies. New courses and seminars in telecommunications and selected areas will be developed. We also plan to develop new curricula in the telecommunications area. Involvement of faculty, students, and researchers from the Columbia University School of Business's program on telecommunications policy, from the Law School, the School of International Affairs, and the School of Journalism should lead to particularly exciting new programs on telecommunications technology and policy.

We estimate that at least 200 undergraduates and an equal number of graduate students from the School of Engineering and Applied Science will be involved in a broad spectrum of research and educational activities once the Center is fully operational. Additional students will be drawn from the other schools noted above.

Industrial involvement with both research and educational activities will be heavily stressed. We are particularly fortunate in our location. Many of the major industrial telecommunications and related research laboratories are in close proximity to Columbia. These include, among others, Bell Laboratories, Bell Communications Research, IBM, RCA, ITT, and Philips Laboratories. Some 1,700 electronics companies, employing 50,000 engineers, are located within 50 miles of our campus. Already numbered among the industrial affiliates of our Center are Bell Communications Research, GTE Laboratories, Philips Laboratories, and Timeplex Corp. AT&T and IBM have provided us with major gifts. For a number of years we have had a close working relationship, through an NSF university-industry cooperative research grant, with a group at IBM Research (Yorktown Heights). Close relationships have also been established with Bell Laboratories, IBM East Fishkill, Codex (a division of Motorola), and a number of other companies. We plan to expand these relationships and seek added financial support from these and other industrial organizations.

Apart from nearness to telecommunications manufacturing, common carrier, and research organizations, our location in the heart of New York City also puts us in close proximity to the largest financial community in the world. These banks and financial institutions are, collectively, among the world's largest users of telecommunications. Merrill Lynch and Citicorp have already provided us with financial support for our activities. We hope to have increased support from these sources in the future.

What means can we use to ensure close cooperation with industry? Over the past few years our weekly seminar series on computer communications networks has regularly attracted speakers and many participants from industry, in addition to our own faculty and students. This series will be expanded to encompass the more multidisciplinary nature of Center activities. We plan to have workshops on a regular basis that are geared to industrial participation. Our graduate courses have always been well attended by part-time students from nearby industry. The new courses and curricula should generate even more interest.

We are currently organizing an adjunct and industrial visitors program through which outstanding engineers and scientists from industry will participate in our research and teaching activities on a regular basis. One possibility, which has been viewed with favor, is to have a weekly graduate-credit seminar with a limited number of students, run jointly by an industrial visitor and a faculty member. A number of such seminars would be organized each semester; they would be expected to develop into cooperative research activities. We also plan to develop short courses for industry in order to provide practicing engineers with continuing education in this rapidly developing field.

Finally, an Industrial Advisory Board made up of executives from a number of leading corporations is being set up. We plan to include on the board a number of government research leaders and representatives of other universities. This board will be expected to provide advice and suggestions as to research direction, industrial involvement, and educational activities. It will also participate in the annual technical review of Center activities by attending research overviews, and by providing names of outstanding engineers and scientists to serve as peer reviewers of annual proposals for research support prepared by Center researchers.

Biotechnology Process Engineering Center

DANIEL I. C. WANG

INTRODUCTION

Fundamental discoveries in the biological sciences during the past 10 years have been truly monumental. Through advances in molecular biology man's ability to manipulate biological activities and properties in both prokaryotic and eukaryotic organisms has created a new engineering field termed "biotechnology." This technology has enabled us to explore the potential impacts of biological systems across a range of applications, including human and animal health, agriculture, chemicals, food, energy, and environment. It has been forecast that by the year 2000 commercial biotechnology markets could reach $40 to $200 billion dollars.

It is important to note that new and fundamental discoveries in molecular biology are emerging on a day-to-day basis. Many of these discoveries provide the enabling technology for new and important commercial products; yet development of the necessary engineering technology has progressed slowly. To overcome this barrier to process and product development, engineering research and training must be accelerated at a rate commensurate with progress in the life sciences.

It is the intention of the Biotechnology Process Engineering Center at Massachusetts Institute of Technology (MIT) to establish well-targeted efforts to advance the research and the training of engineers needed to solve the problems associated with utilization of biotechnology. There is no doubt that biotechnology will become a major manufacturing industry in the very near future. There is, therefore, an urgent need to develop technological concepts to implement this industry. There will also be a need for people to lead and maintain the international competitiveness of this new industry.

It will be an infant industry, ideally suited for creativity and innovation. However, one should recognize that biotechnology will require both broad and cross-disciplinary training. Therefore, it is our goal to train a new breed of professionals through creative interdisciplinary education and research. These professionals will possess the necessary tools, from the molecular sciences as well as from engineering, to be at the crosscutting frontier of the new technology. We also plan to implement our educational and research programs through synergistic and imaginative cross-disciplinary interactions. It is imperative that our programs maintain an active interface with the industrial sectors.

Three MIT departments, and a total of 16 faculty members, are ready to make this major commitment to the Center's education and research programs. The three departments are: (1) Department of Chemical Engineering (11 faculty members); (2) Department of Biology (3 faculty members); and (3) Department of Applied Biological Sciences (2 faculty members).

The selection of these departments to participate in the proposed Biotechnology Process Engineering Center was appropriate for a number of reasons. First, the faculty and departments already have in place coherent educational and research programs directed toward biotechnology. Furthermore, the faculty, as well as the MIT administration, are committed to the development of biotechnology. Research interests of the faculty members participating in this Center are already addressing critical issues in biotechnology. We believe that the 16 faculty members represent the critical mass needed to enable us to execute and implement the programs of the Center. It should be said, however, that we believe opportunities will exist for the future expansion of the Center's activities. There are individual faculty members in other departments, such as Chemistry, Materials Sciences and Engineering, Mechanical Engineering, and the Sloan School of Business Management, who have already expressed an interest in participating in the Center's activities at some point in the future. Collaborations with other universities and institutes are also envisioned.

STRUCTURE, MANAGEMENT, AND PLANNING OF THE CENTER

The overall structure of the Biotechnology Process Engineering Center is shown in Figure 1. The director of the Center will report directly to the dean of engineering. There are four formal committees and programs associated with the Center. These are the Policy Committee, the Operating Committee, the Industrial Advisory Board, and the Industrial Biotechnology Liaison Program. To assist the daily operation of the Center, the

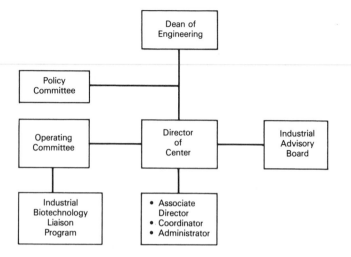

FIGURE 1   Management plan for MIT's Biotechnology Process Engineering Center.

director will have an associate director, and there will be an administrative staff to coordinate the Center's overall activities.

The functions of the different committees should be explained. The Policy Committee consists of the dean of the School of Engineering and the dean of the School of Science, along with the three department heads (Chemical Engineering, Biology, and Applied Biological Sciences). The director of the Center is also a member. The role of the Policy Committee is to ensure the quality and excellence of the Center's activities. In addition, this committee will coordinate institutewide policies for biotechnology in the present, as well as for the future. This committee will also be responsible for the consolidation and allocation of laboratory space needed for the cross-disciplinary programs within the Center. Since participants in the Center programs cut across both departments and schools, the Policy Committee is in an excellent position to formulate optimal policies for joint appointments between departments and schools. Lastly, the Policy Committee will act as an interface with industry with respect to future cooperative programs and future gift programs in support of the Center's activities.

To assist in the Center's overall operations and activities an Operating Committee has been formed. The members of this committee will serve on a two-year rotational basis. Members consist of the director of the Center as chairman, three participating faculty members from the Department of Chemical Engineering, and one member each from the departments of Biology and Applied Biological Sciences. Three members from industry will also serve on this committee. The role of the Operating Committee is to ensure the technical excellence of the programs within the Center. This committee will set priorities for and coordinate both the educational and research activities of the Center. Prioritization could take the form of peer review of existing and future research programs both within the university and with industry. The multidisciplinary background of Operating Committee members ensures that they are well qualified to identify future needs such as new courses, textbooks, faculty, and scientific directions. This committee will also act as the formal link between the Biotechnology Process Engineering Center and the industrial sector, as represented in the advisory and liaison programs. Lastly, this committee will be responsible for relations and interactions with MIT's Interdisciplinary Biotechnology Program, as well as for the student and industrial intern activities of the Center.

To ensure meaningful collaboration and cooperation with the industrial sector, an Industrial Advisory Board has been formed. Members of this board will be senior managers from industry, including the chemical, pharmaceutical, and biotechnology industries. The role of the Industrial Advisory Board is to address the pressing needs of industry with respect to education and research in order to enhance and ensure our international competitiveness. The Industrial Advisory Board will serve two functions: to advise on the present and future activities of the Center relative to industrial needs, and to act as a catalyst for collaboration between the activities of the Center and the industrial sector at large. Lastly, the board will facilitate the identification of technical personnel for liaison between MIT and various private companies.

To instill a more formal industrial collaboration there will be an Industrial Biotechnology Liaison Program. This program will have a broad industrial interface, with no fixed number of companies or participants. The purpose of the program will be to provide technical liaison between this Center and industry in the areas of education and research. This program will be coordinated carefully through the administrative office of the Center. Members of the Industrial Liaison Program will identify the mutual interests as well as the mutual collaborative opportunities of the Center and industry. Furthermore, through this program the sharing of facilities and exchange of personnel can be implemented. It should be noted that this program is to be quite broad in scope and not limited to any one sector of industry. For example, we have identified chemical, pharmaceutical, biotechnology, food, process engineering, instrumentation, and equipment companies as representing the types of industry with which the Center would like to interact through its Industrial Liaison Program.

EDUCATIONAL COMPONENTS

Undergraduate Programs

An overview of the educational programs associated with this Center is shown in Figure 2. A strong educational program will be vital to the success of the Biotechnology Process Engineering Center. It is our belief that entering freshmen must be made aware of potential opportunities in the exploding biotechnology industry. This will be achieved through faculty counseling and special seminars in biotechnology directed at the freshman level.

We plan to have the Department of Chemical Engineering play an important role in undergraduate education associated with the Center. We will not initiate or develop a new degree program, because we believe that a strong chemical engineering base is ideal for subsequent education

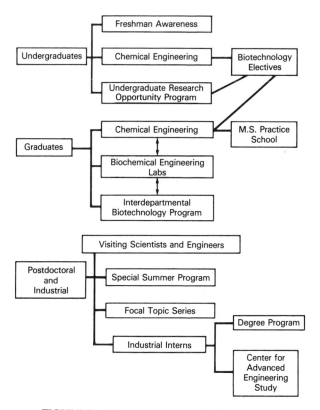

FIGURE 2   Overview of educational programs.

in biotechnology. To instill the needed interdisciplinary perspectives, we envision the use of liberal elective course policies whereby existing biotechnology subjects at MIT can be incorporated into these undergraduate study programs. Faculty participation will play a vital role in counseling and advising undergraduates as to choices regarding graduate studies and industrial opportunities in biotechnology.

Undergraduate research at MIT is an integral part of the overall educational program. MIT has 15 years of experience with a unique Undergraduate Research Opportunities Program (UROP). The institute's UROP office will allow this Center to easily reach the entire undergraduate community. The Center will thus be in an ideal position to offer the undergraduate a wide variety of opportunities to perform research with an interdisciplinary flavor.

Graduate Programs

The graduate educational program associated with this Center will be fulfilled in a number of ways. The Department of Chemical Engineering will have one of the major roles in graduate education. However, rather than initiating a new degree program in chemical engineering, we will first focus on the necessary core subjects in chemical engineering fundamentals. To complement the graduate education associated with the Center, existing electives in biotechnology are ideally suited. At the present time the readily identifiable biotechnology electives include 7 graduate courses in chemical engineering, 6 in biology, 4 in applied biological sciences, and 2 in chemistry. We plan, in the future, to introduce new courses especially addressing advanced principles in biotechnology process engineering, to be presented either singly or jointly with cross-disciplinary relevancy.

In the Department of Applied Biological Sciences, active M.S. and Ph.D. degree programs in biochemical engineering have been in existence for more than 20 years. The education of candidates in these programs is designed to incorporate interdisciplinary skills through courses from the departments of Applied Biological Sciences, Chemical Engineering, and Biology. The doctoral qualifying and written examinations are prepared by members of all three departments. Doctoral Examination Committee members are usually from more than two departments so as to ensure the cross-disciplinary nature of candidates' research. We envision even more interdepartmental interactions in the future arising from the activities of the new Center.

The MIT Interdepartmental Biotechnology Program (IBP) is currently in the initial stages of planning and implementation. However, the formation of the new Center will accelerate the implementation of this program. In 1983 the concept of the IBP was developed to provide, at the doctoral level, educational skills that cross disciplinary boundaries. Thus, the liaison with the Center will strengthen graduate education and research in both activities.

Postdoctoral and Industrial Programs

Postdoctoral training in engineering has not been as common as it has in the sciences. The emergence of biotechnology has begun to change that situation. Recent years have seen an increased interest on the part of engineers in furthering their education in biological fundamentals, and in integrating this knowledge with engineering principles. The proposed Center will play a vital role in coordinating postdoctoral education and training of industrial personnel in biotechnology. This can be achieved in a number of ways. First, the Visiting Scientists and Engineers Program permits industrial and academic personnel to train and study for short or extended periods of time. However, these visiting scientists and engineers are generally not enrolled in the formal degree program.

A second and more formal education and training program offered to industry is through MIT's special summer courses. At MIT, the Office of Summer Sessions offers more than 50 programs per year. Many of the Center's faculty teach these courses, and several of the offered courses are already within the scope of the Biotechnology Process Engineering Center. They include, for example, a special summer course (now in its twenty-fifth year) entitled "Fermentation Technology." Several other courses, such as "Biotechnology: Microbial Principles for Fuels and Chemicals and Ingredients" and "Controlled Drug Release and Delivery," are also offered to the industrial sector as formal training through this special summer program. In the future we are prepared to offer additional courses relevant to biotechnology as special summer courses. Furthermore, we plan to incorporate laboratory techniques in these special training programs for industrial personnel. We also envision special lecture series to be presented at industrial sites—often a practical approach for companies, as more of their personnel are able to participate. Special focal topics and seminars will also be presented at MIT for attendees from the industrial sector. These seminars, presented by participants of this Center as well as other MIT faculty, will also provide the ideal forum for information dissemination and technology exchange.

Formal industrial and/or university internship programs will be established in the future. For example, industrial interns can presently matriculate within a department's degree program at MIT. However, to provide flexibility to industry, internships not associated with degree programs are also possible. This option is ideally suited for our existing Center for

Advanced Engineering Study (CAES). Lastly, having university interns at industrial sites could represent a fruitful educational tool for both student and faculty members of the institute. All of these programs can be readily implemented in the future under the auspices of the Biotechnology Process Engineering Center.

RESEARCH PROGRAMS

Overview and Rationale

Discoveries in molecular biology—especially in the development of genetic engineering—have not only catalyzed interest in biotechnology, but have also provided the scientific basis for a new branch of the biochemical process industry. This new industrial branch, which revolves around recombinant DNA technology, is still in its infancy. There are many products for the human pharmaceutical and animal healthcare markets currently undergoing clinical trials—e.g., the interferons, human and bovine growth hormones, and tissue plasminogen activator (TPA). Other new therapeutic materials, agricultural products, and chemical materials that will be made through a combination of genetic engineering and bioprocessing will have major benefits for mankind, and thus represent major commercial markets. When visualizing these potential benefits and markets, one has to ask: What are the technical barriers preventing commercialization?

Looking further ahead to the second- and third-generation products and processes that will evolve from the new biotechnology industry, two important generalizations can be made. The first is that many important human therapeutic products are proteins with multiple polypeptide chains that are modified by complex biological reactions, often by the addition of complex oligosaccharides. Such modification occurs post-translationally and is required for biological activity and stability. The second generalization is that many of the desired products have low specific activity (i.e., effectiveness per unit weight), and as a consequence will be required in large volumes at low cost. New principles for cost-effective manufacturing processes are required.

The problem has several parts. First, most of the current recombinant-based processes utilize bacteria as a means for production; not only are these processes expensive, but also the bacteria cannot glycosylate or otherwise modify the recombinant protein. Furthermore, when many animal proteins are manufactured in bacteria, they are produced in denatured and/or modified form with decreased biological activity. Second, although fermentation or biosynthesis is the enabling technology, a significant fraction of the manufacturing cost is incurred during product recovery. If we are going to be able to economically synthesize the next generation of products, then new approaches and new technologies must be developed. The establishment of the Biotechnology Process Engineering Center will create a unique locus for collaborative studies between engineers and scientists, focused on breaking new frontiers and developing the fundamental scientific and engineering bases for advanced biochemical manufacturing technologies.

To achieve these objectives the research program in the Center will focus on immediately critical issues in biotechnology. The goal of that research is to explore fundamental principles related to the manufacturing of products from biotechnology. The Center will not neglect concepts for the future that may be long-range in nature. We believe that a cross-disciplinary effort is ideal for deriving maximal and synergistic benefits. Within this research program four generic areas will be addressed (see Figure 3). They are:

- genetics and molecular biology for protein synthesis, processing, and excretion in animal cells and yeasts
- concepts of bioreactor design, scale-up, and operation
- downstream processing for product isolation and purification
- biochemical process systems engineering.

Brief descriptions of each of these research areas are presented below.

Genetics and Molecular Biology for Protein Synthesis, Processing, and Excretion in Animal Cells and Yeast

The newest and perhaps most interesting class of pharmaceutical compounds are human proteins. Members of this class range from hormones

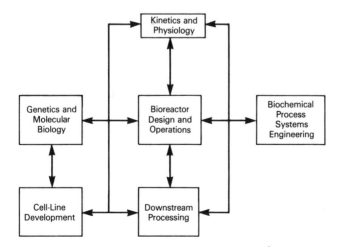

FIGURE 3   MIT Biotechnology Process Engineering Research programs.

(such as human growth hormones) and interferons (such as alpha, beta, and gamma) to specific proteases (such as tissue plasminogen activator) and protease inhibitors (such as alpha 1-antitrypsin, and monoclonal immunoglobins). There are probably a hundred known human proteins that would be routinely used in the treatment of patients if available in pure form and at a reasonable expense. The development of recombinant DNA methodologies has made the isolation of DNA segments encoding these proteins almost routine. Development of technologies to synthesize active proteins, using mammalian cells and yeast engineered by recombinant DNA, is the objective of this segment of the research program.

A number of important problems now restrict the use of recombinant cultured mammalian cells and yeast for industrial production of human (and animal) proteins. The specific problems to be addressed by the Center include:

- production of specific proteins by animal cells
  - vectors for high-level expression
  - control of RNA processing and translation
  - active expression in stationary cells (bioreactor)
- modifications of recombinant derived proteins
  - post-translational events to control protein modifications
  - inter- and intrachain disulfide bonds
  - specific cleavages and additions
- genetic approach controlling protein excretion in yeast
- fundamental understanding of protein misfolding
  - monoclonal antibodies to probe proper folding domain
  - variants with better protein folding ability (downstream processing).

Concepts in Bioreactor Design, Scale-up, and Operation

The bioreactor is the heart of the process in which value is added to the raw material through biosynthesis or biocatalysis. The bioreactor also interfaces with all other aspects of bioprocess development. The productivity of cells results from the cell-line development, and the performance of the bioreactor may define new problems to be solved by genetics and molecular biology. Similarly, the productivity of the cells and the bioreactor design is reflected in the purity and concentration of the product, which determines the difficulty of downstream processing. Lastly, optimization of bioreactor performance and its integration with separation processes require application of process systems engineering.

A coordinated and integrated group of research projects will be carried out. Their selection was based upon a number of considerations. Future production of biologicals, especially animal and human proteins, will require the use of eukaryotic (animal) cells. Although protein production by animal cells has been demonstrated using recombinant DNA and hybridoma technologies, previous industrial exploitation and the existing scientific and technical base are very limited for large-scale culture of animal cells as compared to fermentation of prokaryotes (e.g., bacteria). The overall goals of Center research are therefore (1) to develop fundamental engineering principles for large-scale cell culture, including bioreactor design, scale-up, and operation for animal cells and protein-secreting microorganisms; and (2) to develop strategies and designs which maximize productivity and minimize cost.

With regard to bioreactor systems for animal cells, the following list summarizes the specific research topics that will be investigated:

- characterization of recombinant animal cells
  - –physiology, biochemistry
  - –cellular and intrinsic kinetics
  - –mathematical modeling
- engineering principles
  - –reaction engineering
  - –fluid dynamics
  - –transport phenomena
  - –process control and optimization
- bioreactor design
  - –suspension culture with hollow-fiber perfusion
  - –microcarriers and microencapsulation
  - –novel systems: foams, porous matrix.

Another category of Center research will concern computer control strategies for high-density fermentations using primarily recombinant DNA microorganisms (although the same issues will be of concern with animal cell bioreactors). Process control of feedbatch fermentations will be examined with respect to the effect of control strategy on productivity and product concentration in the bioreactor effluent. Whereas scale-up studies in fermentations have traditionally involved simply increasing scale by an order of magnitude, new methodologies will be examined in which the effect of individual variables (e.g., poor mixing, pH, nutrient concentration, and temperature) are examined with well-characterized, instrumented laboratory bioreactors.

Lastly, the technology for maximizing production of excreted protein products by controlled recycling of cells will be studied with a computer-controlled bioreactor and microfiltration membrane system. This project will be integrated with the more detailed studies of microfiltration of cell suspensions carried out within the downstream processing research group.

Downstream Processing for Product Isolation and Purification

A problem common to all biochemical processes, whether based on fermentation or cell culture technology, is the need to recover the product. In the case of protein production, especially human therapeutic proteins, these products must be recovered in a highly purified form, with the molecule in its proper three-dimensional configuration. The need for extremes in purity, for retention of molecular configuration, and for efficiency in recovery and process scale-up are major challenges facing the bioprocess engineer. As described above, the problem of recovery depends very much on the type of cell and how the bioreactor is designed and operated. Thus, the solution to downstream processing problems will come from collaboration between the biologist and the engineer. A unique aspect of the Center's program on downstream processing research is the close interaction between biologists and engineers, who must continually ask each other: Is this problem best solved through a biological or an engineering approach?

The research will focus on developing broadly applicable, generic solutions to problems of protein recovery, using a variety of different proteins. This is necessary because protein recovery is a multifaceted problem (see Figure 4). Close collaboration between biologists and engineers and interaction with industry will be increasingly necessary to solve new problems today and in the future. Specific areas of Center research in downstream processing are:

- cross-flow membrane filtration
- recovery of insoluble, intracellular proteins
- kinetic approach to adsorption chromatography
- affinity escort chromatography
- high-performance liquid chromatography
- immunoadsorption chromatography
- extraction in biphasic aqueous systems
- protein recovery with reversed micelles
- integration of downstream processing.

A common theme throughout this effort on downstream processing is process integration. We plan to consider each unit operation available and

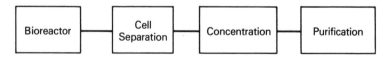

FIGURE 4   Downstream processing.

formulate models that will be used in a complete systems approach to biochemical process development. The development of a systems strategy will greatly enhance our ability to design and operate the most advanced and competitive commercial processes.

Biochemical Process Systems Engineering

The proposed research in this area has been designed in such a way as to achieve the following objectives: (1) use of the fundamental scientific insights developed in the foregoing research efforts to provide systematic engineering approaches and tools for the analysis, synthesis, evaluation, optimization, and control of complete biochemical process flowsheets; and (2) identification of the critical phenomena and/or parameters that may inhibit the industrial realization of new biochemical processing concepts.

The work in this area encompasses three types of research with unified themes.

Conceiving Novel Biochemical Processes The objective here is to guide the systematic generation and evaluation of alternative biochemical pathways for the commercial production of desired chemicals. The value of this work is to be found in the expanded capability of the designer to search through the multitude of biochemical pathways in order to identify the most promising bioprocessing concepts. These in turn will determine the scope of laboratory research and development activities.

Synthesis and Simulation in the Design of Complete Bioprocessing Systems The aim here is to develop systematic procedures for the synthesis of optimum bioreactor configurations, for the synthesis of the best sequence of separations, and for the integration of bioreactor and separation subsystems in order to yield the optimum overall process. Computer-aided simulation capabilities will be developed for the analysis and evaluation of given complete processes. Such analysis will identify critical design and operational parameters leading to process optimization.

Systematic Approaches to Modeling, Analysis, and Control of Biological Processes The two basic objectives here are the development of fundamental insights into the operational characteristics of biological processes, and the design of optimum controllers for efficient operation of such processes. Specific technical issues to be covered are new approaches to bioprocess modeling, synthesis of optimum control strategies, and experimental investigations.

SUMMARY

These four generic areas of research represent a major commitment of intellectual effort to the Biotechnology Process Engineering Center. The research and educational programs, coupled with industrial participation, should result in continuous leadership for the United States in biotechnology manufacturing.

Methods for Ensuring Information and Technology Exchange Among the Centers

CARL W. HALL

INTRODUCTION

In considering possibilities for methods of information and technology exchange among the Engineering Research Centers (ERCs) I will take as my theme the words of a well-known and successful industry research manager, a person whom I was privileged to know: Charles F. Kettering. He said, "When you lock the doors of the laboratory, you keep out more than you keep in." That was a revolutionary view, in comparison to what most industry laboratory and research managers of his day believed.

We need to keep the doors of communication open. The Centers represent a large and long-term investment in engineering research and engineering education. We want to do all that is necessary to ensure that the Centers' activities benefit U.S. engineering schools and serve the national interest. That means finding effective and efficient ways to get research results and innovations in education transferred to users in industry and academia. Although this might be obvious, there could be a temptation to emphasize research and to neglect education. By education I include the whole spectrum from undergraduate education, and perhaps even pre-undergraduate studies, through graduate education to continuing education.

OPTIONS FOR INFORMATION EXCHANGE

It should be emphasized that our plans for establishing additional Centers hinge on our ability to maintain a healthy balance between engineering research project support, support for Engineering Research Centers, and other special engineering project funding. These activities should reinforce each other; consideration needs to be given to how the exchange of information between individuals and the Centers can help accomplish this. The National Science Foundation (NSF) has had considerable experience with the funding and oversight of research centers in diverse fields. These include the National Astronomy Centers, the Materials Research Laboratories, the Submicron Research Center, the National Center for Atmospheric Research, the Regional Instrumentation Centers for Chemistry, Cooperative Experimental Research (Computer) facilities (CERs), and the Industry/University Cooperative Research Centers. Of these, I believe that only the computer division's CERs are on a computer network. The possibility of tying all the ERCs together for exchange of information should be considered.

Center Directors' Meetings

Our experience with other research center operations has convinced us that it is beneficial to center managers and to the NSF to convene periodic meetings of center directors. These are usually held annually, but at the beginning of a program holding such meetings twice a year often proves beneficial. Topics for an Engineering Research Center directors' meeting might include: a report of progress in implementing the Center; the status of industry participation; recruitment plans and discussion of problems associated with building research teams; discussion of education projects aimed at both graduate and undergraduate students; and a range of subjects dealing with the administration of the Centers.

Contract administration, equipment purchase, and maintenance agreements, which have proven to be particularly fruitful areas for collaboration among centers because many of the problems in these categories are common to most center-type operations, might also be discussed. The benefits that can be realized from directorship meetings are well established, and we expect that this will also prove to be the case with the Engineering Research Centers.

The NSF Role: Cooperator and Facilitator

The Foundation's role in ERC directors' meetings would be to act as a facilitator. That is, we are a cooperator in this effort, and we want to assist the Center management as appropriate. We will, of course, be mindful of the adage "Too many cooks can spoil the broth."

It should also be emphasized that the NSF is determined not to micromanage the Centers. The planning principle might be stated as "Give the bird room to fly." Our desire is to create an environment in which NSF is viewed as a constructive cooperator in the effort to achieve the Centers' goals. This attitude will encourage exchange of information and help ensure the success of the Centers.

We will try to avoid procedures and requirements that would cause more paperwork for the Centers. We want to extend the objectives of the Paperwork Reduction Act to our work with the Centers.

With these goals in mind, we are considering the formation of a small technical group within NSF for each of the Centers. Each group would periodically visit the Center it is to monitor. The group would go to the Center with the objective of offering constructive suggestions on the technical aspects of the research program, and helping in the exchange of information.

Through this type of ongoing interaction we expect to accomplish the necessary monitoring without burdening the Centers with a lot of extra data-gathering or reporting requirements. If we do this correctly, the Centers will look forward to the arrival of these teams and all parties will benefit from the interaction.

Computer Networking

In addition to management meetings and periodic visits by our technical teams, we want to explore the options for computer networking and for taking advantage of the availability of supercomputers through the Foundation's Advanced Scientific Computing program. Through computer networks the Centers can benefit from the advantages of electronic mail, electronic bulletin boards, exchange of graphic data, and other capabilities offered by such networks.

I have been told that five of the six ERC awardee institutions participate in BITNET, which is a network of more than 400 university computers, linked via leased telephone lines for exchange of information. Membership in BITNET is free, but new participants are responsible for the cost of both a 9,600-baud leased telephone line to a nearby site and two modems for that line. Thus, BITNET offers the potential for quickly hooking up the eight institutions that comprise the six Centers. However, there are about 60 computer information networks in operation in the United States today, with a wide range of capabilities; so the options are not limited.

The need for a computer network must be completely justified. Questions such as the following must be answered: What are the information needs of the Centers? Who will use the network? What types of messages will be sent over the system? What criteria would be used for accessing the system? Who should be permitted to access the network? Would any special services be offered to companies that contribute funds to the Center or Centers?

The Defense Advanced Research Projects Agency (DARPA) operates the DARPA NET, which is dedicated to researchers working on all aspects of computer science and engineering computer networking. NSF funds the Computer Science NET, or CS NET. About 130 academic institutions are participants in CS NET. This system has been limited to computer science and engineering researchers.

CS NET participants have access to DARPA NET through an arrangement worked out between NSF and DARPA. These systems are worth noting because, under a new NSF–DARPA agreement, it is now possible for NSF grantees in any field of science or engineering to use DARPA NET directly. The criteria for accessing CS NET are being reviewed, and we expect that the Centers will be able to take advantage of CS NET services soon.

The National Science Foundation is also well along with the development of a much more ambitious plan that calls for the establishment of what was referred to as SCIENCENET until recently. Someone has apparently already registered that name, so we are searching for another. I will refer to the system as NSF NET.

NSF NET is an ambitious undertaking. In 1983 scientists and engineers from diverse fields participated in a workshop which focused on courses of action that might be taken to meet the need for computer and network resources in academic science and engineering research. The workshop concluded that there was an immediate need to make supercomputers more available to academic scientists and engineers, and that computer networks are necessary to link researchers to large-scale computing resources and to each other. Efforts to increase the availability and accessibility of supercomputers to engineers may be familiar to many.

I believe that computer networks that provide the user with a wide menu of information transfer alternatives, plus access to a supercomputer, can dramatically enhance the engineering research and education potential in the United States. With a system such as NSF NET the entire United States could be viewed as a single region for research purposes. Given such a setting, in many situations a research colleague or collaborator is only a keyboard away; a researcher can, via the display screen, transmit simultaneous copies of graphics or other work to a number of interested researchers and teachers working on a particular topic. The physical location of a research facility is likely to become much less important. Communicating via a computer network will, I believe, completely revolutionize our thinking on this point.

The goal of NSF NET is to provide a standardized network environment in which users physically remote from supercomputers or other computing resources enjoy levels of service indistinguishable from those of local users. The first phase of NSF NET is thus to make supercomputers quickly accessible to as many users as possible, employing as many existing computer networks as are available.

The second phase will be standardized access. This involves standard gateways that will allow networks with different architectures to interconnect, using standard interfaces and a set of standard protocols to support such user applications as file transfer, interactive graphics, remote terminal access, electronic mail, and remote job entry. Such a network would probably have to employ powerful work stations at the user's site, coupled via NSF NET to supercomputer centers. The potential of NSF NET is great, and it will be a key long-term consideration as we explore options for the Engineering Research Centers.

The Engineering Research Center being established at Columbia University will be pushing the state of the art in telecommunications. This Center will explore the network aspects of integration and will implement the highly flexible network test bed called MAGNET, described in Dr. Schwartz's paper.

Whatever network is adopted for the Centers should be practical, easy to use, and relatively inexpensive. The major objective is to build communication links that will contribute to understanding and that will speed the knowledge process along. A major challenge is to harmonize the networks.

If one speculates a little in this area, it is easy to envision a situation in which an investigator puts a question on the electronic bulletin board and shortly gets an answer from someone he has never met or knew existed. In a real sense, such networks can extend our research horizon, improve productivity in laboratories, and enhance instructional programs across the land. All six of the awardee institutions have considerable experience with information networks. So we are not starting from ground zero in this quest for the best information network.

Other Exchange Mechanisms

We must not limit our thinking to computer networks as a means of information transfer among Centers. Other important mechanisms include:

• People transfer. The most effective means of transferring information is people—whether they be students, faculty, or industry people. An uninhibited flow of people into and out of the Centers must occur.

• Technology transfer (as distinguished from information transfer). Experimental devices and instruments developed by one Center or its collaborators should be made available to others, keeping in mind the importance of recognition of the creator, patent rights, etc. The "NIH"—Not Invented Here—syndrome must be overcome.

• Written transfer (publications). The use of newsletters, project summaries, and electronic and conventional mail can be effective. The possibility of developing new journals—perhaps on computer disks—on various cross-disciplinary engineering subjects should also be considered.

• Verbal transfer (seminars/symposia workshops). We do not expect computer networks ever to replace these important face-to-face discussions.

The networks of exchange probably should not be limited to the established Engineering Research Centers and those to follow. The ERCs should be closely connected with other institutions. For example, many engineering schools do not have extensive research activities, but graduate a large proportion of American engineers; for this reason they are sometimes referred to as predominantly undergraduate institutions. Meeting the overall goals of the ERCs with respect to national competitiveness will require a favorable working relationship of the Centers with some of these institutions. Numerous methods of involving industry people, in both research and teaching, will be tried.

Kettering said, in commenting on which fuel was best for the automobile, "Let the engine decide." In situations involving other institutions and organizations, we should "Let the Centers decide." It is clear that we are going to have to use a variety of mechanisms to extend the benefits of the Engineering Research Centers to engineering schools across America.

BASIC PRINCIPLES

Where does all this leave us? At this juncture it leaves us with more questions than answers. The important thing is that we do not overlook any of the important questions as we move ahead.

1. What are the information and management coordination needs of the Centers?

2. What types of networks and management coordination mechanisms will best meet those needs?

3. Who will use the networks, and who will participate in the management coordinating groups if they are established?

4. What criteria for access will be used, especially for universities and industries that are not participating in the funding of the Centers?

5. What are the best techniques or mechanisms to use in determining potential users of Center research and educational program results?

Emphasis should be given to a point that has occurred to me repeatedly as I have considered the matter of information and technology exchange among the Centers and between the Centers and their participants. It is not a new idea. Justice Oliver Wendell Holmes put his finger on it a long time ago when he said, "Having science in the attic is fine, so long as you remember to use common sense in the living room." The success of the Centers will depend in large measure on the application of a great deal of common sense in their day-to-day operation. All of us—Center management, the NSF, industrial participants, and others who seek to take advantage of the research and educational potential of the Centers—must use common sense unsparingly.

For example, a prime purpose of the Engineering Research Centers is to develop fundamental knowledge that will give U.S. industry an edge in the race for better and improved technologies. If NSF were to attempt to establish safeguards over the transfer of information to protect U.S. interests, there could evolve such a snarl of paperwork that the Centers could be rendered ineffective even before they got started. We are counting on all the participants to use the rule of reason so that U.S. interests are served. To the maximum extent possible, NSF is pledged to avoid issuing guidance papers and other such directives that could impede and frustrate the ERCs instead of helping them to achieve their intended purpose.

CONCLUSION

It would be interesting to contemplate what Charles Kettering might say about the Engineering Research Centers.

• I know that he would be in favor of university-industry cooperation, as he promoted this practice in his own activities.

• I know that he would be in favor of cross-disciplinary research, as he received engineering degrees in two different fields.

• I know that he would favor innovative approaches, as he did when he went against the conventional wisdom in using a small motor to operate the cash register.

• I know that he would urge people to think—an attitude to be encouraged by the Centers. Once when asked to what he attributed his success in innovation, he explained it this way: "As a youth, I had trouble with my eyes [in fact, he stayed out of school a while], so I couldn't spend a lot of time reading books and papers which said a thing couldn't be done." Now I know that he read and studied a lot. What he was really saying was: also THINK and ACT.

• I know that he would favor involving students in real-life situations. He once said, "It's one thing to produce something in the laboratory test tubes and another to manufacture it by the ton."

• I know that he would favor using the experimental approach, as he did when he said "Let the engine decide."

I believe that Charles Kettering would be a strong supporter and salesman for the ERC concept.

The story is told that at one point Kettering had a difficult time getting the production people to accept a fast-drying paint which, he knew, would greatly accelerate the manufacturing of automobiles. He took an important vice-president to lunch. "Now," he said to the vice-president, "if you could have another color of car, what would you select?" "Blue," came the answer. And at a signal the painters painted the V.-P.'s car blue. After a quick lunch they returned to the car, and there it was—beautiful, blue, and dry. Kettering made his sale.

I hope we have "made our sale" of the Engineering Research Center concept. It is an important purchase for the nation to make.

DISCUSSION

There was some discussion of the possibilities for networking and data exchange with respect to the Centers. Dr. Hall noted that each Center will determine its own networking program, but he would expect each Center to involve relevant sectors of industry in the network. Dr. Pipes commented that plans for this are already under way in each ERC; he gave the example of a "dial-up" service at the University of Delaware Center, which, when in place, will make data of all kinds available to participating companies at any time.

New Factors in the Relationship Between Engineering Education and Research

JERRIER A. HADDAD

It is taken as an article of faith that research ensures vitality and competence and thereby improves the pedagogical ability of faculty. However, this faith is not shared by everyone. There are those who subscribe to the "Mr. Chips" school of thought. In their minds, teaching ability is somehow separate and independent from the subject at hand, that is to say, "a good teacher can teach anything." In the engineering area this argument is further complicated by the dichotomy between "practitioners" and "academics." More than any other profession, engineering must rely for its continuing renewal on the 2 percent of its number who fundamentally do not practice, except for whatever engineering research they may do. Especially since World War II, faculty members have increasingly held the Ph.D., and have been selected for tenure only if they could show outstanding research capabilities. There is probably no set of issues that can stir more emotion than these at meetings of university trustees. Discussions about the relationship between research and teaching ability or the difference between the academic and the practitioner have all the elements of an intellectual donnybrook.

Can we strip away the emotional content of the debate and get to the heart of the matter? Most certainly! To begin with, there is simply too much evidence supporting the notion that an engineer or academic who does good research makes a superior teacher. Are there good teachers who do not do research? Certainly! Are there good researchers who are bad teachers? Certainly! How many good researchers are bad teachers? In this day of faculty evaluation by students and tenure procedures that evaluate teaching ability, there are not many. More often than not, student evaluations of teaching ability and administrative evaluations of research ability point to the same people.

The issue of the academic versus the practitioner is getting more complex, however, for a number of reasons:

• Increasingly, practitioners must rely on the latest scientific knowledge to be competitive. This puts the practitioner in the position either of doing engineering research or of being in close touch with researchers. Most researchers who communicate with a range of industrial practitioners are career academics or governmental employees. Industrial researchers are much more constrained.

• Engineering technology is progressing at a very fast rate, both in academe and in industry. Thus, to stay well informed engineers in industry must communicate with academics and vice versa. (Getting out of date is not exclusive to industrial practitioners.)

• Engineering practitioners in government and industry specialize along many dimensions in addition to that of their primary engineering discipline. Their jobs will be in such diverse areas as applied research, product development, manufacturing, manufacturing research, manufacturing engineering, field engineering, engineering or manufacturing operations, service or maintenance, or a host of other engineering specialties. Even these jobs differ substantially in technical content depending on the given industry. This complexity of the engineer's job content makes relating to engineers in faculty positions quite difficult. Engineers in government or industry truly live in different worlds from their colleagues on faculties.

It should come as no surprise that academe and industry are two very different cultures with different values and vastly different practices. It is a matter of some urgency that both groups learn more about each other, become more knowledgeable regarding each other's problems and dependencies, and, especially, learn how best to interact so that each can benefit from the other's empathy as well as its technical contribution.

This is really a very important matter. Academics educate our successors and are the primary source of research that fuels the engineering engine. Practitioners do little research, but do most of the engineering work that fuels our economy, keeps us domestically and internationally competitive, and advances our manufacturing. The engineer in practice gets results in the most scientific manner possible. More often than not, however, project success is attained pragmatically, and, being based on insufficient knowledge, may contain surprises, sometimes of disastrous proportions. Such surprises point the way for further research, and so engineering leads to research just as research leads to engineering.

The problem is to devise means that enable the academic researcher and the industrial practitioner to complement each other best without either having to forsake his own world or invade the other's. Clearly, the Engineering Research Centers (ERCs) were devised as one solution. How will they affect engineering research and education?

THE ERCs' EFFECT ON ACADEMIC RESEARCH AND EDUCATION

ERCs should greatly influence academic research. Industry's heavy participation should help communicate to researchers the problems of execution that stand in industry's way. While many of these problems would have been communicated to the campus in any event, the ERCs will clearly expedite the process and help ensure that the "two culture" syndrome does not slow or block the transmission. Many industry puzzles have stimulated research programs in academe, yielding beneficial results. To the degree that the ERCs can contribute to this process we will all benefit.

Optimistic as we may be about the ERCs, we should not expect them to be a cure-all. To begin with, they can only involve a fraction of the faculty. Those faculty members in fields removed from the focus of an ERC will receive only fleeting benefit from the presence of that ERC on campus. Nonetheless, the values and practices in evidence at the ERC will be communicated through faculty club discussions, luncheon conversations, and cocktail party chitchat.

It is a stated objective of the ERC program to involve both undergraduate and graduate students in the Centers' work. To the degree that this is done, those students will benefit greatly. This is a form of interning. The Committee on the Education and Utilization of the Engineer (CEUE) has concluded that all engineering students should have some form of interning since it has such a positive effect on the student's attitude toward the university experience (NRC, 1985a). Not only does interning bring in a practice component, but it also makes the students see the value of the knowledge they gain from their studies. Interning nurtures personal characteristics that come mainly from experience: positive attitudes, interests, values, needs and motives, and affective skills. These skills are listed as the most important interning goals—even over technical knowledge—by students, graduates, faculty, and supervisors.

It is also an objective of the ERCs that the industry people assigned to them provide a two-way connection to industrial activities, moving campus research results to industry and industrial nonproprietary results to the campus. There is little doubt that ERC's will expedite this two-way communication. However, we should not lose sight of the fact that the industry people assigned will come from companies' applied research sections. Generally speaking, these individuals are quite far removed from the marketplace on the one hand and from the production on the other. In general there has been little difficulty in arranging liaisons between campus and industrial researchers. The problem has been and continues to be putting campus researchers in touch with industrial professionals close to the market or the manufacturing scene. We should not expect the ERCs to have much effect in this regard.

SUPPORT OF THE UNDERGRADUATE SCHOOLS

There is a much more serious problem, however. It is not a problem unique to the ERC program. Rather, it involves the widespread and laudable practice of rewarding already excellent institutions with further opportunities to increase their excellence. It is hard to argue against this practice, and I certainly do not mean to. However, it completely ignores the more than 200 engineering schools that mainly educate undergraduates and that need help perhaps even more than the comparative handful of research institutions. It is a fact that schools that award 14 or fewer Ph.D.s a year award close to half the nation's B.S. engineering degrees (NRC, 1985a).

The CEUE recommends that we invent better ways to support the undergraduate programs in this second tier of schools. The two tiers are a relatively new phenomenon, having come about largely after World War II as a result of contracting from the mission agencies and the newly created National Science Foundation. The largely undergraduate schools seem to be quiet institutions, lacking influence in the technical community and in government and industry. Nonetheless, they are important to the nation and add considerably to the diversity and richness of our engineering education system. It is certainly worthwhile to consider creative ways of improving their situation. A large problem in this regard is how to give the proper support without rewarding mediocrity and encouraging complacency. These schools need help, but we must take care to help in ways that lift the standards and level of education. How to accomplish this is a tough problem that is yet unsolved.

There are state programs of support for undergraduate schools that seem to be working rather well. Consideration could be given to having similar national programs. I will mention two New York State programs that differ in that one is focused on the student while the other is focused on the institution.

The first is called the Tuition Assistance Plan, or TAP. TAP provides assistance to students based on financial need, if they are New York State residents and attend schools in that state. The great merit of this program is that the schools must be attractive to students. Students have a way of picking the best school within the range of their ability to pay. TAP does not attempt to distinguish the relative quality of the various schools, and also leaves untouched the different costs of state-supported and independent schools. Thus, students of a wide range of abilities are able to attend a wide range of schools. One of the great features of higher education in this country is the continuum of quality that is available to students. In my view, we must beware of any scheme, no matter how attractive, that stratifies higher education by means of a bureaucracy. I have much greater faith in the workings of a free marketplace that allows students to pick the programs best suited to their individual needs.

Schools selected by TAP students are free to use the tuition money to do what they deem will make the particular school more attractive to students. As long as the burden of paying all tuition is not placed on the student, tuition costs can rise closer to the tuition the school actually needs in order to attain the excellence of instruction it seeks in the manner it judges best. With accreditation guaranteeing minimum quality, a diversity of schools can best satisfy the nation's needs. If, as a society, we judge that the accreditation minimum should be raised, the matter can be discussed with the Accreditation Board for Engineering and Technology (ABET), which is composed of public-spirited engineers from a cross section of professional societies.

The other assistance plan is called the Bundy Plan. Some years ago, while McGeorge Bundy was president of the Ford Foundation, he was asked by the state of New York to recommend a way to keep alive the colleges in the state. Then as now higher education was having its problems. As implemented, the plan gave "Bundy Money" to degree-granting institutions according to the number and types of degrees they annually granted. In the beginning each bachelor's degree earned $400 for the college, each master's degree $400, and each Ph.D. $2,400. These amounts have been increased from time to time, until for the 1985–1986 academic year they will become $1,500 for the bachelor's degree, $950 for the master's degree, and $4,550 for the Ph.D. In addition, the two-year associate's degree warrants $600.

Here again the attraction of the plan is that the schools must use the funds to continue attracting the students they need for the degrees they want to grant. If standards are lowered to maximize the number of degrees granted, then the most talented students will stay away. If they are lowered significantly, then ABET will withhold accreditation. A powerful motivator is the attractiveness of graduates to graduate schools or the job market. In a free economy you cannot fool the marketplace for long.

A significantly different approach would be to have programs aimed at giving each accredited engineering college at least some research funding. Proposals could be judged by people having no connection with the proposing college. A minimum amount, perhaps based on the student population or faculty size, could be given to support the school's most deserving faculty for their competitive standing at that school, independent of national competition. In doing this we would be encouraging the most talented department to raise its standards, thus fulfilling the goal of improving the preparation of engineers. Further, other faculty members would undoubtedly be stimulated by the competition and seek to improve their research proposals and programs. I know that this flies in the face of the peer review system, which officially ignores the institution. However, our aim is to improve the preparation of our engineers, and to do this we must improve the institutions in the second tier, those educating half of the nation's engineers. If research will improve the pedagogical skill of the faculty, that purpose is just as valid and important to us as the more accepted purpose of adding to fundamental knowledge.

A National Science Foundation program comes quite close to what I have just described. The program, called Research in Undergraduate Institutions, is only a year or two old, and it seems to be successful in many regards. It is designed to give awards to the smaller schools that are predominately undergraduate. "Predominately undergraduate" is in this case defined as granting 20 or fewer Ph.D.s annually in science and engineering. The disappointing thing is that engineering faculties have not responded with as many proposals as the science faculties. I strongly recommend that a survey study determine quickly why this is the case and how it can be remedied.

There is still another approach to distributing the benefits of the ERCs to more colleges than will qualify to host them. As part of their proposals, host institutions could suggest creative ways of involving other, less fortunate colleges—for instance, through faculty summer assignments, sabbatical leaves, student interning arrangements (both graduate and undergraduate), research subcontracts, brainstorming sessions, seminars, and review sessions. Certainly the most appropriate means to distribute benefits will depend on many things, such as area of research, geographic factors, laboratory space and equipment, and the areas of competence of faculty and students. Each situation will be different, and each will require different methods.

NEW FACTORS AFFECTING ENGINEERING EDUCATION

To sum up the considerations involved in the relationship between education and research, it is desirable to list those factors which are either new or have changed in importance in the last few years.

1. The breadth, depth, rate of change, sophistication, and importance of technology and engineering methods in industrial and governmental activities have created a new world for educators to deal with. To design

a curriculum for today's engineering student that is as complete as that of two or three decades ago, and to keep it within four years, is difficult to the point of impossibility.

2. Engineering jobs in industry are highly diverse. The job categories of all employed engineers break down as follows (NRC, 1985b):

Research	4.7%
Development (including design)	27.9%
R&D management	8.7%
Other management	19.3%
Teaching	2.1%
Production or inspection	16.6%
Other (consulting, computing, etc.)	20.7%

Into this broad range of specialties must be factored an industrial specialty and a basic discipline. Educators cannot possibly be expert in all these activities. How to provide an education for these activities on top of an already crowded four-year program is mind boggling. Yet the increasing sophistication and importance of these activities decree that the education system somehow must accommodate them to a greater degree than before.

3. About 200 engineering colleges are predominately undergraduate institutions that produce half the B.S.s in engineering annually. These institutions lack the advantages that research institutions enjoy: world-class faculty, state-of-the-art laboratories and equipment, and the supporting infrastructure that these things bring. If we are to raise the quality and ability of the graduating engineer, we must focus on the graduates of these institutions as well as the research institutions. We cannot and should not aspire to make all engineering colleges into world-class research institutions. However, we cannot stop short of improving the education experience for all engineering students in areas that count most.

4. Engineering in government and industry is becoming increasingly sophisticated in order to compete in an increasingly competitive world. Practitioners need to know the latest research findings, and researchers need to know the obstacles to engineering progress. Industrial concerns, except for the very large and affluent, cannot possibly do research in all the technical areas of importance to them. Consequently, as a nation we should use our government funds and education system to ensure that we appropriately cover the areas of research that are important to our success in the global marketplace.

5. In the effort to hold the undergraduate engineering program to a nominal four years, courses in practical skills have had to all but disappear. The increase of engineering research on our campuses should involve as much student interning as possible so as to expose as many students as

possible to the real world of laboratory work. Such laboratory exposure is considerably better than the standard laboratory course.

6. A serious and perennial problem for faculty is to keep abreast of progress in engineering around the world. This makes faculty contact with practitioners essential. Any program that brings serious practitioners to the campus for technical dialogue is invaluable to the education process. Since an ERC cannot be expected to stimulate this type of interaction outside its technical area, special efforts should be made to introduce the industry people to other elements of the engineering college.

Taken together, these points say that the ERCs represent an idea whose time has come. The ERCs are the first really creative response to a number of interrelated problems. We should labor hard to make them work.

REFERENCES

National Research Council (NRC). 1985a. Engineering education and practice in the United States: Foundations of our techno-economic future. Report of the Committee on the Education and Utilization of the Engineer. Washington, D.C.: National Academy Press.

National Research Council (NRC). 1985b. Engineering employment characteristics. Report of the Panel on Engineering Employment Characteristics, Committee on the Education and Utilization of the Engineer. Washington, D.C.: National Academy Press.

IV

The Future—Challenges and Expectations

Challenges of a Technologically Competitive World: A Vision of the Year 2000

JAMES BRIAN QUINN

The year 2000, which looked so distant for so long, is now practically upon us. In fact, the work the Engineering Research Centers start this year will be exploited mostly after the year 2000. What trends and challenges are likely to continue throughout that time? What are the most likely implications for the Centers? Technology and history are so full of surprises that I will not attempt any precise estimates of future states of the art. Instead, I will attempt some surprise-free comments about the future. One should not be seriously surprised if trends already existing create the results predicted (Kahn and Wiener, 1967). Of course, unexpected major events—a war, political upheaval, or unforeseen accident—could change the picture enormously.

WORLD POPULATION AND WEALTH

The world's population is expected to be about 6.2 billion people in the year 2000, with almost all the growth occurring in developing countries (Figure 1). This growth in population—to 1.4 billion people more than we have today—is greater than the current population of China. Yet this growth is only a point in a continuum toward a likely population of 8 billion people a few decades later. Growth in world gross national product (GNP) has fallen from the annual 5% per year enjoyed through the mid-1970s, yet even the currently expected growth rates of 2.7% to 3.5% (Frisch, 1983) have formidable consequences. By the year 2000 real wealth should be 50% to 66% above 1985 levels. Recent spurts in wealth and productivity gains in the Asian rim, China (which has shown productivity pockets of hunger in the world today. Yet world food production per capita has actually been greater than ever before in both developed and developing countries (Figure 2). The much-maligned "green revolution" has brought important relief to many areas of the world with the development of dwarfed and higher-yield crops, but often at the cost of significantly increased energy and chemical requirements for the land. Diffusion of

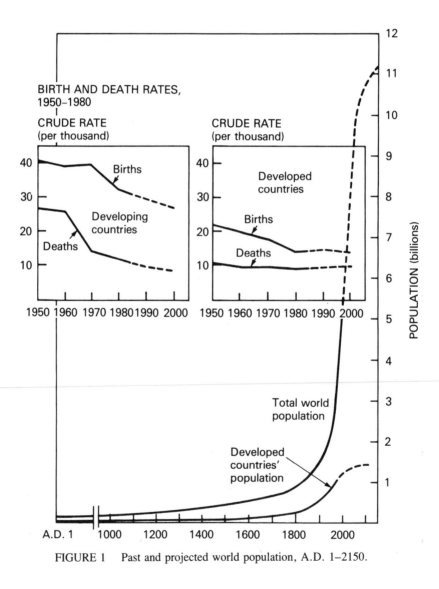

FIGURE 1   Past and projected world population, A.D. 1–2150.

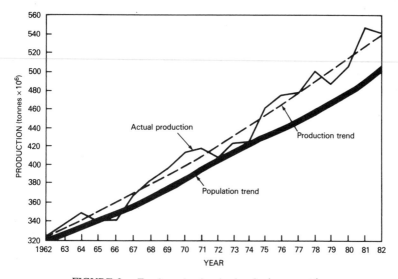

FIGURE 2   Food production in developing countries.

gains of 7% to 8% per year in both agriculture and industry since 1978), and other developing countries suggest the possibility of even higher gains (*The Economist*, 1984). *Worldcasts* and the World Energy Conference estimate a world GNP of about $17.7 trillion (in 1983 dollars) for the year 2000, representing a world market of $7.8 trillion (in 1983 dollars) beyond today's levels (World Bank, 1985). Wealth per capita is expected to rise in real terms from today's $2,000 to between $2,600 and $3,200 by the year 2000 (Frisch, 1983). A key question is whether this wealth will be further concentrated in the developed countries or reasonably distributed among developing countries. Two kinds of technologies—food and energy—will play principal roles in determining this outcome and other competitive patterns in the world.

FOOD AND AGRICULTURAL TECHNOLOGIES

Among the great forces affecting international competition will be food and agriculture technologies. There have often been dire predictions about future world food supplies. Television constantly reminds us of the tragic these technologies will continue to offer productivity increases until the next decade, when advanced biotechnologies are expected to offer even greater potential through higher-yield varieties, improved pest resistance, and better adaptability to saline or low-moisture conditions.

Some experts have estimated that with known technologies the world could feed twice its estimated population of 6.2 billion in the year 2000, and that developing countries could produce two to three times as much food as they do today (Revelle, 1976). For example, pest control could provide enormous gains. Today almost half of all crops produced are destroyed by pests (David Pimentel, personal communication, 1983). Sadly, agricultural technologists know what to do about many of these problems, including the soil destruction that is increasingly moving farms onto ever more marginal lands. Application of known, low-cost technologies such as soil retaining, low tillage, crop cycling, scheduling, land-use planning, and storage could preserve valuable lands, control many pests, and increase usable foods dramatically. Unfortunately, applying more advanced chemical technologies to increase production to the level of developed countries may require capital, energy resources, and technical knowledge that are not always immediately available in the countries that need them most. Getting these resources to where human needs are greatest will be one of the strongest issues, creating potential alliances and conflicts among nations for the next two decades.

TABLE 1   Urban vs. Rural Population Growth in Developing Countries

Income Category	Average Annual Percentage of Population Growth, 1980–2000	
	Urban	Rural
Low income		
Asia (excluding China)	4.2	0.9
India	4.2	1.1
Africa	5.8	1.5
Middle income		
East Asia and Pacific	3.1	0.9
Middle East and North Africa	4.3	1.6
Sub-Sahara	2.9	1.7
Latin America	2.9	0.4
Southern Europe	2.9	−0.2
All developing countries (excluding China)	3.5	1.1

SOURCE: World Bank (1985).

TABLE 2   Trends in Exports from Developing Countries

Commodity	Value of Exports in Billions of Dollars		Annual Growth Rate (Percentage)
	1965	1982	
Manufacturers	7.1	134.6	21.7
Food	13.3	74.8	12.2
Metals and minerals	4.5	26.9	12.6
Fuels	7.3	165.1	23.1

SOURCE: World Bank (1985).

MORE INTENSE LABOR COMPETITION

Certain patterns and consequences of the improvement of food technologies are likely in the near future. While some countries will undoubtedly be plagued by drought and impossible incentive and distribution structures, most countries in the Organization for Economic Cooperation and Development (OECD) will have farm surpluses that are genuine political problems. U.S. farmlands are being abandoned or sold under distressed terms because of high interest rates and the government's refusal to support production at prices higher than those of the world's increasingly competitive markets. U.S. agriculture, which provided the greatest U.S. net export balance—about $20 billion—in 1983, may be on its way to becoming only a "residual source" for world markets, with corresponding negative effects on the U.S. trade balances needed to buy energy and raw materials. Most important, however, in many developing countries about 70 percent of the population has traditionally been employed on farms (*Food Policy*, 1984). Increased agricultural productivity is allowing people to move to cities in unprecedented numbers, creating megalopolises of tens of millions of people, with corresponding huge labor forces that must be employed in nonagricultural tasks (Table 1) (Vining, 1985). These people provide a tremendous pool of cheap labor, which can manufacture with known technologies at very low costs. Cheap labor has begun to change the trade balances of developing countries toward manufacturers (Table 2), and throughout the foreseeable future will create relentless downward pressures on the price of manufactured goods in international trade. Even U.S. agriculture is threatened by imported processed foods (like frozen orange juice from Brazil).

TECHNOLOGY AND CAPITAL TRANSFERS

Each emerging country will urgently seek new ways to form capital through involvement in the more highly value-added industries. U.S.

companies will increasingly seek to produce and source abroad, and capital will certainly be available to those who do so. World capital markets will be ever more closely linked by these ventures through the instant access offered by electronics technologies, and through new worldwide investment and banking structures that exploit these technologies' potentials (*The Economist*, 1985c). With a few exceptions, as in Japan, cost advantages resulting from capital availability will be hard to maintain.

Given the increased rapidity with which technologies have crossed borders (Vernon and Davidson, 1979), permanent technological advantages will be ever more difficult for any single company or country to maintain. The only feasible bases for greater long-term comparative wealth in the United States will be continuous technological and management innovation, more rapid productivity increases in all sectors, and better systems and incentive structures that will encourage U.S. industries to create and adopt new problem solutions. These considerations will be central to the success of the Engineering Research Centers.

ENERGY TECHNOLOGIES

For years the United States based its industrial strength in part on cheap energy and raw materials. Now our relative position with regard to these resources is not so attractive. Although the country enjoys great total resources, these have become marginally more expensive than foreign sources. Despite concerns expressed in the 1970s about limited energy and mineral reserves, the world is slowly recognizing that its ultimately exploitable fossil energy supplies are very extensive, and that its raw materials may be substituted for each other almost without limit, based on their relative prices (Simon, 1981). The Electric Power Research Institute (1981) estimates in its review of world hydrocarbon resources that vast amounts of oil and its substitutes (the equivalent of 7 to 11 trillion barrels, less energy for development and refining) could be available in the very long run with proper combinations of prices and technologies. Although new non-fossil-fuel technologies (and increasing environmental and investment costs for fossil fuels) may mean that most of these hydrocarbon resources are never used, fossil fuels will undoubtedly predominate for the next two decades. The important questions are, at what prices and from what sources?

High Replacement Costs

Although energy costs have temporarily dropped for the United States, this is not true for much of the rest of the world, which has to buy oil in dollars (*The Economist*, 1985a). Developing countries will use more energy per capita as they industrialize. Replacement costs are likely to rise steadily until new synthetic fossil or high-technology approaches are well established. Many replacement sources lie in remote locations and will require investments of many trillions of dollars for development and exploitation over the next 15 years. To the extent that these investments are made in less developed countries, they can provide strong forces driving those nations' economic growth and emergence as attractive world markets and suppliers of other goods.

Few people expect fossil fuels to be as inexpensive as they were in the 1960s; the pressures of politics and replacement costs hold prices up too powerfully. Although we have effected some permanent savings from installed insulation and redesigned engines, it will be interesting to see whether energy growth rates move back toward their pre-1973 values, which were greater than 4 percent, as market forces reassert themselves and energy prices drift toward levels of marginal substitution for other products similar to the levels seen in the early to middle 1970s. The popularity of low-set thermostats, small cars, and slower speeds has already waned rapidly in the United States and OECD countries. A continuing challenge in industrial design will be properly evaluating trade-offs between energy and other costs, including energy-related externalities like acid rain and deposition, polluted groundwaters, and injuries to those most heavily exposed to toxic by-products of energy production and use.

Other Developing Alternatives

By the year 2000 the world will probably have proof of several other large-scale systems offering truly permanent energy access. At Creys Malville, France should have proved continuous breeder reactor operations—if not their economics—on a commercial scale. Although formidable technical problems remain, U.S., Japanese, European, and Russian fusion power programs still seek to surpass Lawson's criterion (energy break-even) within the next decade (Clarke, 1981). The constantly improving field of solar voltaics—a young $180-million business in 1984 (*The Economist*, 1985b)—is another developing alternative. But neither of these will significantly affect energy supplies by the year 2000. Once proved at commercial scales, however, these technologies could offer a long-term prospect, characterized by relatively stable energy costs and fewer environmental problems. More important, they could redefine the very nature, scale, location, and availability of the raw material resources of the world—and thus the longer-term wealth potentials of many now-developing areas.

NEW STRATEGIES FOR THE AMERICAN ECONOMY

Assuming that major trends in these two most important technology areas—foods and energy—develop in the noncatastrophic fashion suggested, what is a likely scenario for U.S. and world industry over the next several decades? While we must assume that the United States and other advanced countries will be increasingly dominated by their service sectors (Figure 3), we must also remember that "services" include many high-technology industries that do not happen to produce a tangible "product": airlines, utilities, communications, retailing, wholesaling, healthcare, banking, insurance, financial services, and others that are very technology-intensive and need continual infusions of engineering science and expertise.

It is difficult to maintain reasonable trade balances, however, solely by exporting services. There will probably be strong pressures to maintain at least a 20 percent employment presence in manufacturing. Solely for reasons of national security, it seems likely that at least viable steel, chemicals, ground transport, aircraft, electronics, domestic energy, and ship-building capabilities must be maintained, by government subsidies if necessary (Quinn, 1983). Other industries that will have to remain internationally competitive must squarely face the problems outlined above: how to compete with some companies (like the Japanese firms) that may have half the capital costs, with others (like those in developing countries) that have a tenth or twentieth of the labor costs, and with still others that have especially low-cost raw materials in addition to low labor costs.

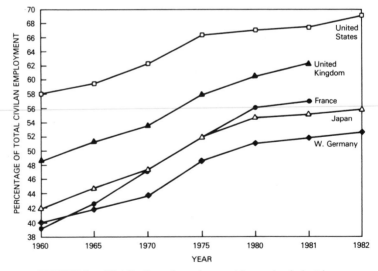

FIGURE 3   Distribution of employment in service industries.

There are few easy answers. Strategies must fall back on what the United States can do best: exploit its extraordinarily rich and varied scientific base; get closer to its own customers in the largest, wealthiest market in the world; relate sciences, technologies, and customer needs to the search for new solutions with higher total value added; and exploit the country's entrepreneurial capabilities and flexible capital structures, which have been the envy of the world. All these strategies require continuous innovation, not just in products, processes, and system technologies, but also in the use of smaller, more flexible organizations and more imaginative management concepts.

ELECTRONICS AND COMMUNICATIONS

Many opportunities will arise from electronics, the most powerful single technology of the current era, and from biological technologies, which will offer a wide range of new solutions for agriculture, human healthcare, environmental improvement, chemical processes, and even energy production. Many important dimensions and trends within the electronics and communications technologies have been well documented. However, it is their interface with other technologies and their use in entirely new system solutions that will present some of the most fascinating horizons of the next 15 years. What are some of the likely effects on industry structures, competitiveness, and management?

The demand for electronics functions has been growing continuously and exponentially for several decades, and it is expected to grow another 100-fold in the next decade. When one asks executives or investors how they would like their company to be in an industry with such a growth rate, they exhibit a mild excitement. Oddly enough, virtually all companies and institutions have the opportunity to share in this growth rate, because it is the use of these technologies that will expand so rapidly in the next decade. Almost everyone is a potential applier of the technology, whether in their travels or at home, in factories or government offices, in retail shops or on farms, in research centers or educational institutions, in healthcare facilities or places of entertainment. The benefits of electronics will accrue most notably to those who apply electronics, not to semiconductor manufacturers, as is often implied in discussions of world trade advantages. Each capability of the technology opens its own particular opportunities.

Communications Bandwidth

Bandwidth, the amount of data that can be carried over a single link per second, has been growing continually and exponentially for decades

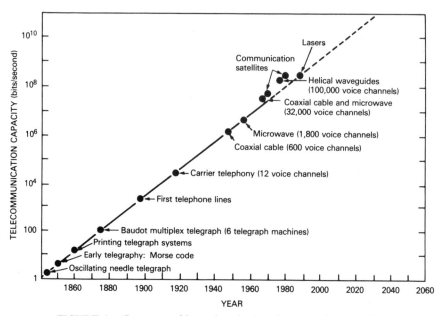

FIGURE 4   Sequence of inventions in the telecommunications field.

(Figure 4). Today advanced laser optics systems in commercial use can transmit approximately one-half billion bits per second—the information equivalent of the words in 100 sizable books. In the laboratory, researchers can now transmit approximately one trillion bits per second—the equivalent of the words in 200,000 books. At this rate lasers could transmit all of the word information in the books of the Library of Congress in less than half an hour. The theoretical potential of light-frequency lasers is still approximately two orders of magnitude beyond the current laboratory art, and significant progress toward that goal should be expected by the year 2000.

Will we use information in the same way when it can be transmitted commercially at these rates? Probably not! Demands will expand markedly. To use this capability effectively will require whole new industry infrastructures. Already major new endeavors have arisen in information packing, mass storage, fiber optics, light-frequency modulation, rapid remote sensing, light-frequency recording and playback, solid-state laser components, light-activated computer devices, and computer security techniques. Other completely new consumer systems (instant remote imaging), institutional systems (medical diagnostic and surgical techniques), communications services (local area networks and data systems), input-output devices (voice and on-line sensing), and linkage systems (electronic jails) seem to pop forth daily to use these huge bandwidths.

Density of Components

The density of components has continued to grow rapidly, although at a slower rate of change than in the early 1970s. Meindl and others have suggested that component densities close to one billion components per chip are possible by the year 2000. Others believe this may be conservative. In any event, such capabilities immediately suggest the potential of very powerful (several picosecond) computers in minuscule packages, fractions of inches in dimension. Of course, along with this extraordinary power and small size come fascinating challenges for software and firmware to program the chips and organize input data so their capacity can be realized. The full power of such devices cannot be exploited without extensive real-time sensing capabilities, high-bandwidth transmission capabilities, and software concepts (like those of artificial intelligence) that decrease the distance from the central processor to its most remote memory sources.

A few chips can give each appliance, factory tool, vehicle, school, office, clinic, home, supermarket, and corner repair shop a computing power that would have been unimaginable in 1970. Given the creative ways people have used increased computer power in the past, one can only wonder what further applications will be commonplace 15 years from now. Electronic capabilities will undoubtedly restructure virtually every institution's research, producing, financing, distribution, servicing, purchasing, and marketing systems. These changes will be at least as important as lowered production costs for most companies. Some examples are discussed below.

Electronics Costs

Historically, semiconductor costs have dropped 25 percent to 30 percent for every doubling of volume; capability has become ever cheaper as power has grown. However, the costs of developing and making the first new chip have ballooned from a few hundred thousand dollars for large (LSI) chips in the late 1970s to tens of millions for very large (VLSI) chips in recent years (Figure 5). Today there are two different views about future cost patterns. Some sophisticated companies say they have found ways of developing and introducing new complex chips that will not inflate future costs. Others say that future generations of chips will take hundreds of millions of dollars to develop and introduce.

Once a production line is set up and debugged, however, it is nearly totally automated. Marginal materials and labor costs approach zero; semiconductor costs become determined by error costs and yields, with prices

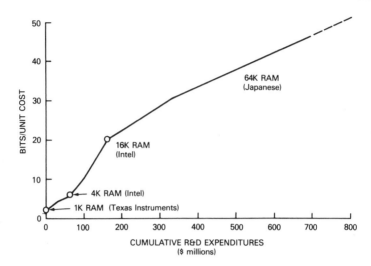

FIGURE 5   Growth of total R&D costs for improvements in semiconductor units of memory. SOURCE: McKinsey & Co.

following low-margin commodity patterns. Few doubt that very dense, powerful, reliable chips will be available in high volumes at low cost in the late 1990s. Still, their use in devices will require entrepreneurial imagination, rapid and flexible production and marketing techniques, and extraordinary attention to quality production and the software without which the chips would be useless. These are familiar problems of mass production, and should be an arena where the United States can compete well if corporate incentive and promotion systems are adjusted to attract and reward well-trained people for quality manufacturing.

As important as computers are, they are increasingly likely to become commodity items, except for those used in extremely advanced laboratory or military applications. If patterns seen in other fields obtain, a few firms with great depth in the technology itself, significant production expertise, and strong distribution capabilities will dominate these commodity markets. The use of computer technologies will be much more crucial to profitability and the generation of wealth in a society than the actual production of semiconductor chips and the off-white boxes that contain them.

Storage Capabilities

Electronic storage capabilities and costs have also improved exponentially over the last several decades. Today all the words in all the (non-duplicate) books in the Library of Congress (some calculate about a quadrillion bits of information) could be stored in an incredibly small space. For example, if laser disks now in the laboratory can store 4 trillion bits of information, only 250 to 1,000 would be required to store the amount of information in the Library of Congress. Beyond today's capabilities lie potentials of another order, in atomic storage—which has essentially unlimited capabilities. To use these capabilities will require a fascinating merger of atomic physics, molecular biology, and chemicals and electronics research: a stimulating set of challenges for a new laboratory system.

Since none of the above subsystems in electronics and communications will approach its theoretical limits for some time, these major technologies should continue to improve in all their important dimensions throughout most of the next decade. What do we do with these technologies as they approach zero marginal cost, operate at the speed of light, occupy almost no space, are able to store infinite information, are immensely reliable, and demonstrate flexibility beyond belief?

Automation and Employment

Many have made the dire prediction that electronics will perform all jobs; hence employment, incomes, and demand will disappear. This has certainly not been the pattern of the past. All studies to date show that electronics has actually increased total employment substantially. In fact, our service-dominated economy today would be impossible without electronics. There is the obviously substantial employment in computer and associated products industries, and service sectors like banking, insurance, air transportation, hospitals, libraries, travel services, education, communications systems, entertainment, government services, and the military would grind to a halt without electronics. All such services seem more likely to expand than contract their employment during the next decade.

Although one cannot predict precisely what new products will spring from the imaginations of inventors and entrepreneurs, they are certain to occur. In the mid-1960s few would have thought that the American home of the 1980s would be a combination of supermarket, movie house, video arcade, short-order restaurant, discotheque, computing center, and automated heating and plumbing establishment. We are now poised to move beyond mere comfort and entertainment to other basic needs like personal medical diagnoses, self-education, home employment, security services, child-monitoring, emergency health services, automated household repairs, transportation services, electronic banking and shopping, and so on ad infinitum.

Farm and ranch homes can adopt electronics technologies in even more spectacular ways to help with hard chores. Automated plant nurseries, and animal- and plant-feeding, irrigation, planting, monitoring, spraying, harvesting, testing, and product-classifying systems already exist. Among the more interesting applications are an automated sheep-shearing system developed in Australia and a chicken-deboning system created for U.S. restaurants.

Further potentials are limited only by human imagination. There should certainly be unbounded opportunities until well beyond the year 2000.

THE FLEXIBLY AUTOMATED FACTORY

Much has been written about the impact of electronics on each type of institution mentioned above. However, the flexibly automated factory illustrates many of the greatest potential impacts of electronics on management and competitiveness (Jelinek and Goldhar, 1984). In such a factory items can be produced in essentially any sequence without incurring substantial extra setup costs. (Each machine is relatively indifferent as to whether it produces 100 of the same item or 100 different items in sequence.) The setups are all preprogrammed. As setup time approaches zero, the Economic Lot Quantity (ELQ) for production approaches one.

This transition shifts economies away from certain well-known "economies of scale" toward "economies of scope." Because producers suffer no additional costs for producing variety, they benefit by absorbing software and hardware investments over the greatest possible range in the marketplace. This means that a flexibly automated producer should compete in as many niches as possible within relevant markets. An experience curve for this family of products replaces the individual skills and learning curves of workers as the variable cost most influencing total costs. Costs decrease most as one improves the software and machine relationships in producing the highest volume of products within the design range of the system (Talaysum et al., 1984).

Customer Orientation

From the viewpoint of marketing it becomes essential that the factory be as closely connected as possible to customers. Ordering systems could electronically link individual customers by computer directly with the plant's production planning system. This would minimize finished stock inventory costs and allow greater market competitiveness by ensuring that the customer gets the precise product and delivery desired. As completely flexible automation is approached, variety in the marketplace not only does not cost the producer more, it adds value for individual customers. On the other hand, since the plant approaches having all costs fixed (other than for materials and parts), unit costs become very volume-sensitive.

Hence, serving each customer and market niche and maintaining continuing market relationships are more important than ever. Since other flexibly automated factories can approach the same cost characteristics, competition moves from traditional cost modes toward more understanding of customer needs and desires, and enhancing product quality and services accordingly. These concepts pose real challenges in understanding for engineering research centers and engineering schools.

It should become possible to produce products of nearly perfect quality each time. This should lower quality costs by decreasing reruns and warranties. Since process control is in software and hardware rather than in labor skills, many products may be produced equally effectively anywhere in the world, thus opening new world markets for a company's product lines. Electronic bandwidths and communications can allow these remote plants to be controlled and monitored with whatever detail is necessary. Consequently, managers can delegate with greater confidence. The technology thus permits highly decentralized organizational structures operating close to customers. Unfortunately, the danger also exists that managements may use the technology to centralize structures and thus drive out the very people and attitudes necessary to design and make the systems work right in the first place.

Lower Costs

Although costs for some pieces of equipment and for software may rise, many other investment costs can be lowered for a variety of reasons. With zero setup costs, a given piece or grouping of equipment can offer higher capacity on the same plant floor space. This should decrease space expenditures, as should the minimizing of personal facilities, heating and lighting requirements, and so on. Investments should also be lowered by the decrease in work-in-progress because of automatic transfers among work stations. Inventories may be further decreased by coordinated (or just-in-time) inventory control systems worked out with suppliers. To realize the full benefits of flexible automation, better supplier relationships become crucial, and entirely new supplier strategies may be necessary. Suppliers must be allowed to invest in these same advanced technologies. This probably means fewer, more sophisticated suppliers per manufacturer, longer-term contracts to justify automation expenditures, and closer relations with suppliers to ensure that quality and delivery specifications are met in all cases (Goldhar and Burnham, 1983).

More dramatically, however, manufacturing labor costs decrease drastically as a percentage of total costs. Automated factories in the United States should begin to approach the unit labor costs of less developed countries. This could assist the resurgence of manufacturing here and in other advanced countries. It may also allow the development of world-class manufacturing facilities in developing countries, but these countries will be relatively hindered by their lack of sophisticated supplier and communications networks. Optimal locations will be increasingly determined by supplier availability and minimized inventory costs, rather than labor cost differentials. All the above points demand longer planning horizons for companies than were required in the past. They also require a global approach to product and manufacturing strategies.

Implementation

Flexible automation requires very careful strategies to implement its complex capabilities within the factory and with necessary supplier and marketing links. To date the approach usually taken is to develop clusters of machines affecting single parts or subassemblies ("automated cells") that slowly bring key operations under control. This allows the company to gain needed experience before linking the whole system into an integrated network and organization. Interestingly, the people who have been "staff" in the past now become "line" personnel. The old bull-of-the-woods supervisor is no longer relevant. Programmers and maintenance people become the core of the operating force, with the remaining work force consisting of a small group of unskilled laborers who punch buttons, watch meters, and sweep floors. As automation occurs the changeover to this new kind of work force requires careful planning, and great care in the retraining of people to minimize personal and organizational distress.

Smaller-Scale, Flexible Operations

A continuing trend in industrial organization appears to be toward smaller-scale, more adaptive operations. For reasons of motivation, cost control, and flexibility, many companies (especially in innovative industries) are trying to keep the number of personnel at individual locations below 500.

Even in large-scale, continuous process industries, the minimills of the steel industry suggest the economies that may be available by decreasing fixed plant overheads and adopting alternate technologies as conditions change. Sociological trends indicate a continuing preference for more individualized items in the consumer trades. These changes, and the need to get ever closer to industrial customers' specific requirements, increasingly seem to be making flexible manufacturing systems more cost-effective than fixed-position automation for specific situations.

Many studies suggest that better understanding of customers is one of the most powerful competitive advantages American producers could have, but one that they have often overlooked in their rush to achieve greater efficiency. Many new forms of flexible design groups—"skunk works," venture teams, and partnering, for example—that allow continuous feedback from customers and active involvement of workers have become essential for rapid and effective innovation. Similarly, Engineering Research Center solutions will have to keep intimately in touch with changing market, organizational, and process needs of users. Not maintaining this orientation properly has been the bane of European industrial research institutes for decades.

HEALTHCARE COSTS AND THE FUTURE OF INDUSTRY

In the last few years certain nonproduction costs of industry have begun to soar. Health costs have doubled as a percentage of GNP since 1960 (Figure 6). Healthcare costs, now included in fringe benefits of major

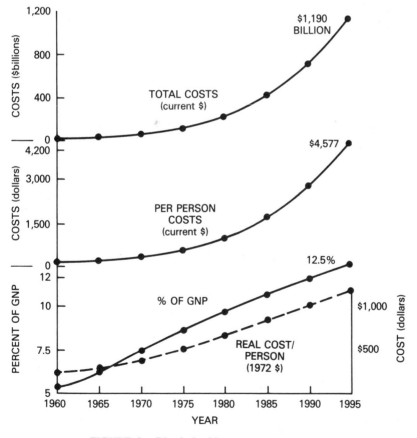

FIGURE 6   Rise in healthcare costs since 1960.

companies, today often exceed the companies' total profit figure. And in some cases work-related health hazards or health liability claims can become large enough to bankrupt major companies (like Johns-Manville and the owners of the Three Mile Island facility).

Carcinogens, Mutagens, and Litigation

Since virtually any chemical substance can be proved carcinogenic or mutagenic if given in massive doses to test animals, there will be increasing tendencies to look to employers to compensate for real or imagined injuries on the job, from products produced, or from wastes dispersed into the environment. Because nearly all products will face such problems, the Engineering Research Centers have a genuine opportunity to help find more benign solutions, where costs can be efficiently absorbed across a broad spectrum of industry. If solutions are not found to these haunting problems, our litigious society can easily close down the very industrial complexes that have provided such great service, power, and wealth in the past. In economic terms, these problems will be at least as pressing as those of production cost containment and quality improvement within the next 15 years.

Leaks into the aquifers of Florida and the Silicon Valley have proved that the economies of entire areas can be destroyed by small-scale chemical intrusions into water supplies. Increasing evidence is accumulating on the health effects of atmospheric pollutants, including carcinogenic particles, leads, aerosols, and acidity. Maps of cancer incidence suggest that living around industrialized cities seems to be dangerous to human health. Complicating action is the long delay between an initial environmental insult and its identification as a recognizable cause of disease. Thirty years elapsed before a correlation was observed between men's smoking and cases of lung cancer, with the pattern tragically repeated later for women.

Similarly, there are many products in use now—hair dyes, drugs, flavor enhancers, paints, component and process chemicals—whose long-term effects cannot be evaluated yet. The businesses that produce these products in all good faith today may be bankrupted years from now in order to pay the unforeseen costs of present decisions. Significant work is needed over the next decade and a half to develop better mechanisms for measuring pollutants, for reducing health risks, and for fairly sharing the natural risks of modern life. Using electronics capabilities, advanced monitoring and sensing systems can be built directly into production processes. In addition, health management on the employer's premises can be improved by automated tests that can be made unobtrusively. The Engineering Research Centers may offer a new vehicle for employers with similar problems to engage in joint research projects, and perhaps ultimately in joint waste-disposal efforts, to improve their practices and lower their (and the nation's) health costs.

National Healthcare Costs

Annual U.S. healthcare costs will probably exceed $1.2 trillion in the mid-1990s, creating enormous national overheads and great challenges for new technologies and systems to contain these costs. New technologies will increasingly allow people to live for months or years through situations they could not have survived a few years ago. As previously deadly diseases are increasingly eradicated, an ever greater percentage of the population will reach late retirement ages. The margins of our productive sectors will have to be expanded to cover these increased national overheads, despite the crushing effects of low-cost foreign competition. This may in fact be the greatest of all engineering challenges. Without appropriate solutions, our society will have to make some difficult choices about who lives on, or be bankrupted by its successes in healthcare.

BIOTECHNOLOGY

Fortunately, technology has provided a dramatic new capability that may balance some important negative trends. Biotechnology has revolutionized medical research and is on its way to revolutionizing healthcare, foods production, chemicals production, and waste processing, possibly ameliorating many of the problems mentioned above. Bacteria have already proved to be remarkably helpful in cleaning up many environmental insults, and will doubtless be more so when they are deliberately bred for this purpose. Similarly, as research develops new knowledge about the natural chemical protective agents of plants, animals, and humans, it should be possible to prevent and cure many diseases that have been intractable in the past.

Harnessing the diagnostic potential of genetic engineering has already led researchers to claim that more progress has been made in cancer research during the last two years than in all preceding history. If the genes that produce antibodies (or other defensive agents) for living systems can be found, cloned, and switched on, experts believe they may be able to produce highly specific entities to attack almost any human, animal, or plant disease diagnosed. When combined with better sensing, monitoring, and early diagnostic capabilities at job locations and disposal points, they should offer much-enhanced future prospects for environmental and human health improvement.

Genetic techniques should also have a major impact on production processes. Agricultural researchers anticipate locating the genes that affect specific inherited plant traits, and then developing plants or seeds with these desired characteristics, thus short-circuiting the years of selective breeding needed to achieve similar results. Plants with desired traits (like bulkier tomatoes, corn with stronger stalks, easier-to-harvest fruits, and even nuttier-tasting wheats and corns for breads) can readily be conceived. Before the year 2000, many researchers also expect to create some grains and legumes that can fix their own nitrogen fertilizers, as the soya plant does today. Although extraordinarily difficult to achieve, such a development could have dramatic effects on less developed countries' food production and energy imports. By the same date, however, and with higher probability, genetically engineered vaccines and hormones should vastly improve animal husbandry and prevent some deadly livestock diseases, like hoof-and-mouth disease and shipping fever.

In the chemical industry many claim that biotechnology will allow smaller-scale, less energy-using, and less waste-producing plants for many products. Virtually any petrochemical derivative or organic chemical is potentially producible by biotechnology. Fructose, amino acids, and analgesics are at the top of the list for large-scale biotechnology operations. But "bugs"—or their brethren, yeasts and cells—are also able to produce indigo dyes, cleaners for river barges, safe noncorrosive substitutes for road salt, and effective leachates for processing minerals such as copper, sulfur, and uranium. Biotechnologies are young, and the above examples only suggest the range of their possible applications. Their effect on the scale, diversity, and location of agriculture and industry is likely to be profound. Surprises will abound as technologists rapidly expand our understanding of the possibilities and limits of biological systems that have been mutating since life began.

PROSPECTS AND CONCLUSIONS

Technological progress over the next 15 years offers great hope for a better world future. The prospects I have alluded to represent minimal advances. Surely inventions and discoveries will unleash new, brighter potentials. We simply do not know what these are now.

To realize these potentials takes political foresight and wisdom we have not always been blessed with. Food and energy developments require massive investments on a worldwide scale. These will only occur if nations see their interests as intertwined, rather than polarized by religious or political ideologies. Populations can be controlled, fed, and made wealthy, but only by nations willing to innovate, to make social investments, and to educate rather than build monuments, perpetuate myths, and create war machines.

Our traditions of independence, entrepreneurship, freedom, and individual rewards have made the United States the greatest technological innovator in history. These forces—supported by an expanding base of scientific understanding—should continue to serve private needs well. But for each pound of goods we produce for our wealthier society, the Law of Conservation of Matter says we must also ultimately produce a pound of waste. Managing the by-products of progress will require new infrastructures, public investments, and planning on a scale we have never before achieved. Business and government will have to work hand in hand in developing new mechanisms that will deal with these needs as well as our private enterprise system has dealt with our traditional product and service needs. The Engineering Research Centers are an encouraging development in attacking some important problems, and we trust that their successful launching will lead to the kinds of technological solutions this country so genuinely needs by the year 2000.

REFERENCES

Clarke, J. F. 1981. An interpretive overview of the United States magnetic fusion program. Proceedings of IEEE 69(8) (August):869–884.

The Economist. 1984. China notes that Marx is dead. Vol. 293 (7373/7374) (Dec. 22, 1984):56.

The Economist. 1985a. Oil and the dollar: the odd couple. Vol. 294 (7376) (Jan. 12, 1985):56–57.

The Economist. 1985b. Solar power: a sunrise industry. Vol. 294 (7385) (Mar. 16, 1985):97.

The Economist. 1985c. International investment banking: a survey. Vol. 294 (7385) (Mar. 16, 1985):1–88.

Electric Power Research Institute (EPRI). 1981. A Review of World Hydrocarbon Resources Assessments. EPRI EA 2658. Palo Alto, Calif.

Food Policy. 1984. Ten years after the World Food Conference. Vol. 9(4) (Nov.):278 et seq.

Frisch, J. R., ed. 1983. Energy 2000–2020: World Prospects and Regional Stresses. Report of the World Energy Conference. London: Graham & Trotman.

Goldhar, J., and D. Burnham. 1983. Concept of the manufacturing system: present and future approaches. Pp. 92–104 in U.S. Leadership in Manufacturing. National Academy of Engineering. Washington, D.C.: National Academy Press.

Jelinek, M., and J. Goldhar. 1984. The strategic implications of the factory of the future. Sloan Management Review. Summer, 1984.

Kahn, H., and A. Wiener. 1967. The Year 2000: A Framework for Speculation. New York: Macmillan.

Quinn, J. B. 1983. Overview of current status of U.S. manufacturing. Pp. 8–52 in U.S. Leadership in Manufacturing. National Academy of Engineering. Washington, D.C.: National Academy Press.

Revelle, R. 1976. The resources available for agriculture. In Foods and Agriculture. Scientific American Books. San Francisco: W. H. Freeman.

Simon, J. 1981. The Ultimate Resource. Princeton: Princeton University Press.

Talaysum, A., M. Hassan, D. Wisnosky, and J. Goldhar. 1984. Scale vs. scope: the long run economies of the CIM/FMS Factory. Chicago, Ill.: Illinois Institute of Technology. Monograph.

Vernon, R., and W. Davidson. 1979. Foreign Production of Technology Intensive Products. Washington, D.C.: National Science Foundation.

Vining, D. R. 1985. The growth of core regions in the Third World. Scientific American 252(4):42–49.

World Bank. 1985. World Development Report 1984. New York, N.Y.

Goals and Needs of U.S. Industry in a Technologically Competitive World

ARDEN L. BEMENT, JR.

INTRODUCTION

Today the United States is being strongly challenged by its international trading partners for world markets. We feel deep concern at the erosion of our great basic industries, such as steel, automobiles, and other heavy manufacturing. These industries have been the backbone of our past economic strength. We are also concerned that many of our recently formed high-technology industries are being placed in the category of "endangered species" by the targeting practices of our world trading partners. We now find ourselves at a crossroads, and must decide whether our current business strategies, institutions, structures, and laws will continue to sustain our high standard of living. As Simon Ramo (1984) has pointed out, "the international race for technological superiority is as ferocious as any cold war battle, and it is fundamental to deterring a hot war. To win can enable a nation to be master of its fate while also enjoying the fruits of superiority in technology. To lose badly can be a catastrophe."

The government is now calling on the scientific and engineering communities to respond to global competition, to make use of emerging technologies to create new products, processes, and management systems. The establishment of the Engineering Research Centers is a major element of this national response.

To envision the goals and needs of U.S. industry through the year 2000 is both a presumptuous and a hopeless task, since in retrospect many of the interesting technologies developed during the past 15 years represent significant discontinuities from the past. By any measure the pace of technology development is accelerating. I find it hard to ponder what further acceleration could occur as our thinking processes are improved with the aid of advanced computers and software. Predictions even out to 15 years must be seen as highly speculative and probably useless.

Nevertheless, there are persistent trends that will nurture the new growth industries. As John Naisbitt (1982) observed, we are restructuring our society to emphasize information over industry, decentralizational over centralization, telecommunications over printing, entrepreneurial activities over managerial ones, and an integrated global economy over a national one.

The 1980s will be a period of uncertainty, during which people will have numerous options and will exercise their options with greater intelligence and creativity. There will be increasing challenges for large companies to be more forward-looking in analyzing the future and in effecting changes. They will have to develop stronger insights into selecting potentially successful technologies and into helping their businesses adapt.

Before looking to the future of American industry, however, I believe the past and present should be put in perspective. In spite of the apparent demise of some of our major industries, as a nation we have been successful in developing new industries over a sustained period of time. Some of the trends we have seen include the following:

- the creation of new jobs at a phenomenal rate—nearly twice the rate of Japan
- a steady and continuing increase in manufacturing productivity
- an industrial output that, as a percentage of gross domestic product, has been remarkably stable over the past 30 years, averaging about 24 percent in spite of two major wars, oil crises, at least two major recessions, and a determined antigrowth, antitechnology movement.

A great deal of credit for these achievements must be given to the skill of our work force and to the successes of our management methods, entrepreneurial vitality, willingness to take risks, and technological contributions to productivity. However, global competition now compels us to improve our performance further—dramatically in most instances—at all stages of technological and business development.

To excel in this competition we must change our attitudes and outlook. We can no longer assume that "what's good for the United States is good for the world." Also, we must change our engineering philosophy from one that teaches "If it ain't broke, don't fix it" to one that teaches "If it's working, even well, then improve it."

FORCES IN OPPOSITION

While exponential changes are taking place in technology over time, much slower changes are occurring in world cultures, social habits and traditions, government institutions, and global policies. In fact it has been the strong desire to preserve social traditions and institutions that has caused technological revolutions to be spread over many decades, even after industrial expertise and commercial feasibility have been well established. Therefore, one can expect that most of the goals and needs of U.S. industry through the year 2000 will be driven by technological revolutions that are already in progress and in various stages of maturity. The more prominent include microelectronics, telecommunications, biotechnology, medical diagnostics and implants, and synthetic materials.

There is growing concern that the broadening gaps in educational levels among U.S. citizens will result in a widening social stratification as the information revolution occurs. The ability to access and manipulate information will be a key survival skill in tomorrow's society, and lacking it could constitute a formidable barrier to upward mobility.

The accelerating pace of technological change will also bring about new issues, concerns, frameworks, and challenges to which global policies and international relationships must adjust. International tensions, such as we are experiencing now in our trade relations with Japan, could intensify as nations accelerate their efforts to capitalize on new technologies, and especially if leadership for managing change is lacking. Ruben Mettler (1982) points out that "all nations have an essential stake in a more unified, open and balanced world economy . . . A healthy world economy that depends on expanding trade won't just happen, we have to make it happen by an increasing effort to cooperate, to resolve our differences constructively . . . and to reduce trade barriers even when it hurts . . . We need new leadership, new strategies, and perhaps new institutions and structures that will enable us to leapfrog across the dismal swamp in which we find ourselves."

U.S. industry is being challenged more now than ever before in the face of growing international competition, not only to plan strategically, but also to manage strategically. Yet the effectiveness of such strategic plans and actions depends critically on industry's ability to reliably forecast future changes in technology and the environment.

Nearly all the environmental factors to which industry must respond are subject to dramatic change. Some forces are dynamically opposed, so that the possible outcomes of an event can be dramatically different depending on which force should dominate. Consider, for example, the following forces in opposition.

First, students of the Kondratieff "long-wave" theory of economic activity argue that as a result of growing world overcapacity we are nearing the end of a long-wave cycle, which will bring about sharper depressions, higher unemployment, shrinking profit margins, more burdensome debt loads, and hyperinflation. Other modelers argue that the current rate of new job creation in the United States, spurred on by new technologies and thriving entrepreneurship, will be sustained and will counter the next long-wave downturn, bringing about serious labor shortages in some regions of the United States over the next decade. Which model is a planner to believe and a manager to act on?

Second, in the coming years there will be a general leveling of technological competence among the highly industrialized countries. An ever-increasing fraction of technological advances will occur outside of the United States. The United States will be required to adopt a fast-follower rather than a leadership strategy in some technologies, not only out of necessity, but also out of economic advantage. Furthermore, many multinational corporations will be marketing to the more than 600 million people of the highly industrialized countries. In contrast to these trends, however, there is growing pressure on the United States government to protect employment by means of industrial policy, to control the flow of technological information, to protect emerging technological industries, and to pursue a policy which advocates United States domination in all of the sciences and technologies, under the assumption that it might even be possible to do so.

Third, many major technological industries in the United States are seeking partners in other countries in order to achieve a broader base for capital investment, technology inputs, skilled labor, and market access. Examples of such consortium strategies can be found in the automotive, aircraft, nuclear, communications, and electronics industries. In opposition to these trends the industrialized countries are attempting to achieve greater national control of new technological enterprises through targeting practices, the erection of nontariff trade barriers, and the nationalization and subsidization of industries. In the United States, the dismantling of major corporations by applying antitrust legislation, which was created for an isolated domestic economy, seems also to be counter to the growing internationalization of competitive forces.

Fourth, worldwide telecommunications systems linked to teleports and backbone connecting networks linked to major finance and industrial centers are emerging. These systems will further internationalize the conduct of world banking, trading in world stock exchanges, and marketing and distribution. More than ever before money flow will be equated to information flow, and investors, the ultimate owners of production capacity, will become more highly distributed around the globe. Opposing these developments are the growing controversies among nations over privacy of information, access to and rights to monitor information flows across borders, and government control of the technical means of information transfer.

These examples represent only a few of the socioeconomic and political forces that can dramatically alter the environment for technological change and industrial growth in the future. These forces do not represent new challenges to the engineering community, but they most likely will become more critical determinants of how an industrial enterprise flourishes in an increasingly competitive world.

Fortunately, some of the emerging technologies may offer new pathways around some of these barriers. For example, they will permit past patterns of doing things, but in new, unconventional ways. They will provide new options for improving life-styles through added convenience, while still preserving social habits and traditions. They can make complexity more intelligible to all by using high technology to make products and services simpler to use. These technologies can bring new flexibilities to managing change. They can open up additional opportunities for wealth while at the same time conserving existing wealth. For these reasons I believe it is critically important for the newly selected Engineering Research Centers to include in their curricula the study of how technological changes in their fields should be managed in the face of prevailing social, economic, and political forces.

U.S. INDUSTRY IN TRANSITION

The law of dynamic competitive advantage requires that American industry constantly introduce new products, processes, and services if it is to remain at a high level. This will require either devising new technologies or using existing technologies in innovative ways.

I believe the information revolution has the greatest potential for changing the course of industry in the next 15 years, primarily because it is already doing so. New information technologies are already linking the factory with the office and the home, and are bringing about changes in the industrial value chain.

The value chain in this context consists of those activities that add value in the industrial enterprise: materials and services; product design and development; production; distribution, marketing, and sales; and after-market operations. Information technology is dramatically changing the relationships among the links of this chain.

First, information technology is making the interfaces among the links more transparent and superfluous. For example, product and manufacturing engineering are already being merged into single engineering units in many manufacturing divisions. Common data bases are also being designed to accommodate information flow all along the value chain from order entry to production scheduling, factory routing, inventory control, and final customer billing.

Second, information technology in the form of direct terminal linkups is extending the value chain to include suppliers at the leading edge and users at the trailing edge. These extensions of the value chain to suppliers and customers are stretching around the world to facilitate world sourcing, marketing, and distribution strategies.

Third, information technology in the form of privately owned local and long-distance telecommunications networks is permitting a greater number of bypasses, or shortcuts, through the value chain. Therefore, it is becoming more cost-effective to allocate noncritical operations to external suppliers and service companies, to distribute knowledge-worker assignments throughout a company, and to implement nationwide and worldwide production and coproduction strategies.

United States industry is already exploiting economies of scope in addition to the previously achieved economies of scale. As a result, service and production industries are becoming more interdependent. Also, bypasses are being sought in marketing, transportation, distribution, and warehousing to force greater convergence of manufacturing costs and final retail costs. For example, the increasing use of telemail ordering from the home is already bypassing the distribution link.

Also, integrated transportation companies are emerging. Through the use of distributed information networks they can select the optimal delivery method, use storage buffers along the delivery chain to best advantage, and provide continuous traceability of the transported items. These changes, plus the full implementation of ''just-in-time'' inventory management by United States industries, will make it increasingly cost-effective for more foreign companies to locate their production facilities in the United States in order to sell to United States markets.

THE FACTORY OF THE FUTURE

The factory of the future is not a new concept; it has been evolving for more than 30 years. It is generally imagined to contain highly integrated manufacturing cells consisting of machine tools, robots, automated materials handling systems, distributed sensors, and a number of digital controllers. In general, these cells would be controlled by a model-driven computer-aided design and manufacturing (CAD/CAM) data base. Additional control programs would be introduced into the cell to provide diagnostic functions that anticipate breakdowns and provide cell control throughout upset conditions, with or without human intervention. Ideally, material stock and information would be fed into flexible manufacturing cells and parts would come out. Such highly tuned cells are considered the key to attaining increased productivity and quality, reduced lead time for producing a new product, improved reliability of production, and reduced manufacturing costs. To be complete, however, the factory of the future must be a fully integrated system rather than just islands of automation.

Integrating the factory of the future will depend on the following kinds of control:

- adaptive control for machining cells
- real-time control of material and information on the factory floor
- production planning control, including scheduling, inventory control, material requirements planning, and capacity adjustment.

Although several plants around the world approximate the factory of the future, most of these currently face major barriers to implementing a fully integrated, computer-based control system. These barriers include: the cost and complexity of available software; the lack of interfacing standards; the lack of appropriate control algorithms and models for system integration; and the lack of expert systems for optimal scheduling and routing.

For a number of reasons, then, fully automated factories are likely to be the exception rather than the rule even by the year 2000. In industries that have frequent model changes, the time and cost necessary to program new software for such changes may be prohibitive. In some plant layouts the manufacturing cell operators, assisted by data displays and decision aids, may be able to control production more economically and with a greater dynamic span of control than a hierarchical computer control system might. Finally, in some cases full automation may not be warranted because it will not contribute sufficiently to the value of the product to justify the investment.

In the future some emerging technologies, many of which are already in use, will add substantial value to manufacturing operations:

- Net and near-net shape fabrication methods will reduce materials use and minimize metal removal operations.
- Improved cutting tools, tool wear sensors, adaptive grinders, and improved abrasives will greatly increase the speed of metal removal operations.
- Advanced lasers will speed up drilling, cutting, and welding and joining operations.
- Plasma deposition and buildup processes will replace some mechanical assembly operations with chemical assembly procedures.
- Advanced surface modification techniques using physical and chem-

ical vapor deposition, ion implantation, and directed energy beam annealing will impart better wear, abrasion, and corrosion resistance.

• Improved computer methods of process modeling and simulation will speed up process design and automation.

• National data bases on the properties of materials, based on fundamental behavior models, will greatly reduce the time required to test and evaluate the use of new materials.

• Automatic tape-laying machines and advanced polymers will make possible the manufacture of high-volume composite parts.

• Highly agile, coordinated robots will be increasingly used in assembly operations.

Moreover, the cost structure for manufacturing operations could change dramatically over the next 15 years for small and medium-sized companies. Because of the prohibitive costs to these companies of hiring the talent needed to revamp their manufacturing technology, "specialty houses" are being established to respond to their demand for technical help. In the future such houses may offer not only engineering consulting services, but also control, communications, and inspection modules; software; equipment rebuilding services; machine tool rentals; and data-base management services. Through these services some manufacturers may opt for a higher ratio of variable to fixed costs, so as to respond more quickly to changes in the business cycle and in technology. Furthermore, as steel-collar workers displace blue-collar workers, direct labor may become more widely regarded as a fixed cost.

THE STEEL INDUSTRY OF THE FUTURE

The steel industry is in a state of ferment, in spite of the restructuring, downsizing, and refocusing of large, integrated steel mills going on today. New technologies are emerging that will enormously improve quality, productivity, and product performance while reducing energy and capital costs.

New information technologies, which can improve the understanding of steelmaking processes and help eliminate processing defects, can greatly improve quality, productivity, and production costs. The steel industry and the federal government have set a standard for cooperation in establishing programs in advanced sensor development. These programs promise to provide sensors that will withstand hostile environments, to make possible the continuous analysis of liquid steel chemistry by laser spectroscopy, and to permit monitoring of temperature distributions throughout large, hot bodies.

Microprocessors are also being applied to real-time process analysis and control so as to improve production yields and product compliance with engineering standards. The use of x-ray tomography can permit better dimensional control of mill products, and reduce the amount of excess steel given away to the customer through lack of control. New, in-line nondestructive characterization methods, which can detect defects in incandescent steel during production, will eliminate the need to cool down steel at intermediate breakdown stages for cold inspection.

New concepts in steelmaking under pilot development around the world are signaling future dramatic changes in the industry. One potential breakthrough is the substitution of coal for coke in the direct reduction of iron ore. The development of improved refractories that are less reactive with liquid steel, as well as improved pouring methods, advances in deoxidation practice, and more extensive use of vacuum degassing and ladle processing, will lead to cleaner steels. These steels will have a lower content of sulfur and other undesirable residual impurities, lower retained oxygen content, and improved inclusion control.

The steel industry has already achieved major advances in productivity and cost reduction by installing continuous casting facilities. It appears that gains from future developments will be even greater. The continuous casting of thin slabs and strip, which will eliminate the need for primary breakdown mills, hot strip mills, and reheat furnaces, will result in further dramatic reductions in capital and operating costs.

Improvements in the chemical homogeneity, microstructural refinement, and porosity control of these near-net shape products will be provided by continuous magnetic stirring and rapid cooling during the solidification process.

The development of dual-phase steels, which has enabled significant weight reductions in automobiles, has been a major success story for the steel industry. However, the full potential for strength improvements and property uniformities needed by the automotive industry for improved formatility is only now being made possible by the installation of continuous heat-treating lines.

Finally, parallel advancements in electrogalvanizing technology, to provide improved laminated and alloyed zinc coatings for corrosion protection, will greatly extend warranty times against cosmetic damage and coating perforations.

I see the next 15 years as extremely challenging for the steel industry—certainly a time in which the industry can demonstrate to the nation that technological revolutions also come to mature, basic industries.

CONCLUSION

In looking at the future of U. S. industry I tend to be optimistic. The United States is still the strongest nation in technology in the world, and we are getting better at using this technological strength competitively. As Ruben Mettler (1984) observed, ''Global competition compels all of us to improve our performance in all aspects of our businesses. This includes making use of all that technology can bring to our products, processes and management. The challenge is not whether to optimize technology but how to develop and select what's best for our purposes, how to control the cost of using it, and how to finance it, all the while earning enough profit to continue to invest and compete in world markets on a sustained basis.''

A major part of this challenge is to remain aware of technological developments around the world. Our universities represent the best means for doing this. Our great research universities combine the functions of education and basic research. They have long been lodestones for the best scientific and engineering students, faculty, and researchers from around the world. As a result, our opportunities for exchanging ideas and being exposed to the world's technologies are greatly enhanced. These opportunities should be nurtured rather than restricted.

The number of models for university-industry interactions have proliferated in recent years, easing the connection between the industrial researcher and the university investigator. While the Ministry for International Trade and Industry (MITI) develops the national strategies for targeting technologies for economic growth in Japan, the United States has already decentralized this process. Most state development offices are preparing regional targeting strategies with the close participation of business and university leaders. To add to our present Silicon Valley, they are actively planning the architectures for biotechnology, polymer, microelectronic, and intelligent manufacturing valleys, using a great diversity of institutional models for technology development, transfer, and reduction to practice.

The Engineering Research Centers sponsored by the National Science Foundation are centers of excellence that can provide impetus to our national engineering research. Most important, these Centers have a responsibility to provide leadership in developing the new curricula that will educate future engineers who can translate our visions into reality.

Notwithstanding the progress already made, the nation still has a major task ahead—that of reequipping our university engineering research laboratories. It is not enough for our universities to model industrial processes and manufacturing operations with computers or simple prototypes. They must also have facilities of a scale sufficient to support the development of advanced industrial process equipment, machine tools, metalworking equipment, control systems, instrumentation, and software.

In my university experience I have sensed how the excitement and challenge of electronic devices, microprocessors, advanced sensors, robots, control systems, and artificial intelligence have contributed to attracting top students to engineering. If our public and private sectors cooperatively support and sustain the enthusiasm of this talent, our nation will go a long way toward meeting its goals and needs in a technologically competitive world.

REFERENCES

Mettler, R. F. 1982. New private initiatives for expanded world trade. Address before the Japan Society, New York, June 23, 1982.

Mettler, R. F. 1984. Charles M. Schwab Memorial Lecture. Address before the American Iron and Steel Institute, New York, May 23, 1984.

Naisbitt, J. 1982. Megatrends: Ten New Directions Transforming Our Lives. New York: Warner Books.

Ramo, S. 1984. U.S. technology policy: an engineer's view. National Academy of Engineering. The Bridge 14, no. 3 (Fall 1984):5.

A Mature but Rejuvenating Industry: Expectations Regarding the Engineering Research Centers

W. DALE COMPTON

The question of the proper relationship between engineering practice and research has been effectively addressed in a number of papers. I have a somewhat similar conflict in concepts to discuss—namely, the expectations of an industry that is at once mature and rejuvenating. After all, maturity means "having attained the normal peak of natural growth and development," while rejuvenation refers to change.

So I am discussing the rejuvenation of something that is mature: elements of industry that are trying to return to their adolescence, in a sense, and to recapture a greater flexibility and a greater capability for innovation. It is in this context that one may ask, what do these industries expect to gain from the Engineering Research Centers (ERCs)?

It is useful to reiterate a few points made in other papers. While it is difficult to generalize about any large segment of an industrial complex, we can find some common traits among a number of our mature industries, including the automotive industry.

First, mature industries are experiencing increased competition from overseas suppliers. The reasons, as Professor Quinn's paper points out, are relative labor costs, the relative value of currencies, tax policies of various governments, and even the targeting of markets by other governments.

A second characteristic of mature industries is the growing need to meet demands on the part of customers for improved product quality. The reason? The customer simply will not accept poor-quality products, let alone shoddy products. Perhaps a more important—certainly an equally important—reason is that producing a high-quality product costs less than repairing a poorly manufactured product. This relates directly to the total cost of manufacture.

The third common characteristic is an increased use of technology as a tool for improving competitiveness. The reason? Technology must be used to offset some of the local advantages of overseas competitors. There are few alternatives, and furthermore, through technology the U.S. manufacturer may be able to offer a variety of products that can more effectively compete in the marketplace.

How are the mature industries attacking these problems? How are they achieving cost reductions? Basically, every aspect of the business is being examined. New management tools are being used, and as I have noted, technology is increasingly being used to help reduce costs. Quality is being upgraded. We now understand far better than before that a product must be designed to meet the customer's needs. It must be designed so that it can be manufactured; and it must be well manufactured. Therefore, the entire process from product conception to final manufacture must be understood to be an integrated system.

Flexibility is being emphasized. Many of our mature industries have large capital facilities. The steel, aluminum, glass, automotive, chemical, and aircraft industries are all examples. We are learning that facilities must be designed to accommodate change at a more rapid rate than we ever experienced before. Today it costs the automotive companies between $700 million and $1 billion to build a new engine plant. This is a sizable percentage of the $2 to $3 billion it costs them to create an all-new vehicle. We simply have to build flexibility into our facilities so that those investments have a longer period of use.

It is appropriate to ask how an Engineering Research Center—no matter how big, no matter how good—can help with the efforts toward cost reduction, quality, and flexibility?

First and foremost, ERCs can offer faculty and students an understanding of the total system and its complexity, from product conception through design and development to final manufacture. Furthermore, they can provide a broad understanding of productivity and how it translates into competitiveness; they can demonstrate that productivity is important in all phases of the process, not just in the final stage called manufacturing. The ERCs provide an opportunity for an in-depth look into the total system. As Roland Schmitt has noted, they provide that laboratory experience for engineering students that has been missing for far too long.

Second, through their research the ERCs can furnish industry with improved general techniques and tools—tools to handle, to manipulate, and to control large and very complex systems.

By way of example, consider the case of a large company that may be carrying an inventory exceeding $1 billion. With the cost of money today, that inventory costs around $100 million a year to maintain. Inventory needs are determined by a large number of factors—e.g., the options that one offers in the product, the reliability of the supply base for materials, and the distance between the supplier and the factory that uses those materials.

To elaborate on the point about options: if you look at an automotive assembly plant, the normal line produces about 60 vehicles per hour; that translates into roughly 400,000 vehicles a year. Thus, one of the Ford Escort assembly plants could build roughly 1.2 million vehicles over three years. With the option content that is currently available for the Escort, that plant could operate for nearly three years and never build two identical vehicles. A natural question arises: How much do those options actually cost? The simple answer is that we are not certain. We simply do not have adequate tools to determine the value of eliminating a particular option or the cost of adding another one. We have some general guidelines, but we really do not have a sufficiently quantitative description of the system to be able to offer that kind of analysis.

Another example of how the ERCs can be invaluable in the area of tools and techniques concerns the technology of robot installation, as Dr. Hackwood emphasizes in her paper. We regularly make decisions as to whether certain welding processes are going to include robots. Now, some design changes are made in automotive vehicles each year. Those changes may require modifications in the assembly process. If the assembly line has a robot in the line at a point at which the line must be changed, it frequently costs more to move the robot than it did to buy it in the first place. Such factors may be critical in deciding the level of automation that is to be introduced. Cost trade-offs have to be made.

These examples emphasize the need for better tools, for better models, and for better simulation techniques for designing the product and manufacturing it. We welcome the trend that we see toward revamping the industrial engineering curricula in this country. The new emphasis on modern manufacturing can make great contributions in training people to help attack these problems.

I hope that the list of topics generated for the 1987 ERC program announcement will contain a number of generic issues that are directly relevant to our so-called "mature" industries. I also hope that we will be generous in our interpretation of relevance, and not make too great a distinction between mature and emerging industries and their relationships to the ERCs.

To illustrate this need for applying general criteria, I wonder how many of the six currently funded ERCs would be readily identified as relevant to the mature industries. Probably not many, and yet they all are. From an automotive point of view, the composites being studied at Delaware

are the materials of the future, not just for hang-on panels, but for structural items. The systems work at Maryland is of direct relevance and of great importance to the understanding and control of our total systems. The effort on integrated circuits manufacturing automation under way at Santa Barbara is of direct interest. The automotive industry will continue to be one of the very largest users of integrated circuits. We design our own circuits; we have to know how they are going to be manufactured.

The computer-based intelligent manufacturing systems work at Purdue is of obvious importance. The effort on networks at Columbia has long-term implications for our industry. As a multinational company, Ford has a communication problem that is immense, particularly as we move toward an all-electronic system and away from a paper system. Finally, the bio-technology effort at MIT can impact the development of new fuels, new materials, new adhesives, and so forth.

Thus, in making those lists I hope the National Science Foundation will be careful about compartmentalization. Relevance is sometimes difficult to gauge.

We should not expect the ERCs to solve specific, immediate problems. That is industry's task. But the ERCs can help create a new state of mind in students—a new outlook and a new approach—so that they will be better able to solve those problems when they join us.

Rejuvenation is a traumatic experience, but for some of us in the mature industries the alternative is even less attractive. As our mature industries strive to become more competitive, we need new employees who have experienced some of the aspects of a rejuvenated engineering education and a rejuvenated research experience. That, it seems to me, is what the ERCs are all about.

A Growth Industry: Expectations Regarding the Engineering Research Centers

The Semiconductor Research Corporation (SRC) welcomes the establishment of the Engineering Research Centers (ERCs) by the National Science Foundation (NSF). These Centers offer the potential for strengthening the engineering capabilities of the United States and enhancing the competitive position of this country in important segments of industry. No industry is more aware of the need for strengthening its competitive position than the integrated circuit industry represented by the SRC. Through the SRC, the integrated circuit industry has established "centers of excellence," with some similarities to ERCs. At the same time, those universities that have been selected to operate an ERC have been given a unique opportunity. If they do not use this opportunity to develop improved institutional environments for applied research, we will all lose.

The functions and complexity of integrated circuits continue to increase, and are the key elements for systems that will allow us to understand, manage, and control information and activity in many areas of human endeavor. For this reason we tend to equate the integrated circuit industry with the information technology industry, and to believe that United States success in integrated circuits will be central in future economic growth. The integrated circuit industry looks to the universities for three important resources: well-trained graduates, new ideas, and high-quality research results. University research centers that have a concentration of effort, experience, facilities, and skills are the primary source of such resources. These centers supplement much larger research efforts in industry and government laboratories that are generally more strongly focused on goals and products.

OPPORTUNITIES FOR ERC UNIVERSITIES

There are three overlapping opportunity areas that a university operating an ERC should address: motivation, management, and growth.

The motivation opportunity is related to the structure of the university and its reward system. Most universities are now structured around discipline-oriented departments, and a faculty member's stature and rewards are strongly focused on personal achievements as determined by peers within the discipline. However, progress in engineering research often demands strong interdisciplinary collaboration and the subordination of individual goals to those of a team. The careers of some faculty members have been adversely affected when they gave priority to such collaborations. Such experiences predestine a research center to a limited existence. A new approach to motivating the best faculty to participate in Center research may do more than anything else to make an ERC successful.

A second opportunity relates to the stature of a Center within the university. Research centers in general are all too often supported by department chairmen only until they in some way threaten the departmental control of funding or staff. Then the support may erode, and the center ends up with inadequate staff or resources, and ultimately disappears. There must be strong motivations for continued departmental support of research centers, encouraged by university administration. Despite the fact that universities have been widely recognized as sources of expert consultation on management, it is generally accepted that universities are poorly managed.

Historically the university has been a loose confederation of scholars. In the modern world, universities have become big businesses. The annual research expenditures at MIT are about $200 million, and there are more than 50 U.S. universities at which they exceed $5 million. For most of these funds the university has the contractual responsibility. It fulfills the responsibility by delegating the responsibility to faculty members. Responsibility for performance and deliverables are placed completely in the hands of the performer, with minimum, if any, oversight. Results are predictable

Contractual requirements are often neglected and research commitments left unfulfilled. On the other hand, the quality of the research product depends largely on the freedom of the individual faculty member. If increased management were to decrease the quality of research, it would be the equivalent of shooting oneself in the foot. The Engineering Research Centers will depend on good management to be productive. That follows from the problem-solving nature of engineering research. The opportunity is for universities, perhaps calling on some of their expert consultants, to find mechanisms for better management of their enterprises while preserving their research quality. Fifty percent of academic research expenditures in engineering are concentrated in the 14 top schools. In research related to integrated circuits, the SRC identifies 6 schools in the top tier of research capability. One objective of the SRC, and of NSF in establishing the ERCs, is to elevate the research productivity of additional universities. This is a challenging goal, almost too challenging. The research environment that attracts excellent faculty and the best graduate students to a given school evolves over a long time, and requires both strong technical leadership and a committed institutional structure. The universities at which ERCs are being located should, in a few years, be among the top universities in their technical areas as well as in broader areas of engineering research.

CENTER OPERATIONS

The specific attributes that a research center requires to meet the expectations of its constituency include a unique purpose, goal-oriented research, problem identification, effective dissemination activity, and good management.

The university model for a research center is often to define an area of interest and to gather faculty participation within this area. The specific agenda then is defined by the interests of the faculty and is often a relabeling of ongoing research. The SRC model for a research center includes defining a unique goal or purpose, defining a research vehicle for demonstrating progress, establishing the relevance of various research tasks, and an effective management structure. Under the contractual aegis of the center the SRC at times supports research unrelated to the focus of the center, but this is separately reviewed and evaluated. The uniqueness of the center goal recognizes that increased benefits will result if different centers work on different things, and that there are an adequate number of macroengineering problems to provide a unique problem for each of the centers the SRC can establish. In our view, the goal of an Engineering Research Center should be more than that of defining an area of research. For example, the SRC-Cornell Center for Microscience and Technology has as its goal the demonstration of 0.25-micron silicon technology in a configuration compatible with a 16-megabit dynamic random-access memory (DRAM).

The goal orientation of the research is perhaps what should distinguish an ERC from materials or science research centers that are more fundamentally oriented. The goals for the various research tasks performed within the ERC and their relevance to the Center goal should be clear. The goals should be assigned target dates, as in an industry research project. This arrangement would provide excellent training for students by providing realistic discipline.

Problem identification is a core concern of Engineering Research Centers. History is rife with solutions to nonexistent problems, and we are increasingly aware that much research merely repeats prior research. Lacking omnipotent information bases, the Engineering Research Centers must build strong external constituencies for their research in order to remain relevant and useful. The opportunity here is to define a role for universities in engineering research that does not compete directly with that of industry, but which contributes in important ways to industry's generic research base. At the same time, the Engineering Research Centers must deviate from the traditional university mode of addressing collections of small problems to focus on the larger, more complex, and more important problems of today's industry. To take an example from SRC research, a large problem is the efficient and rapid transfer of data among the parts of a multimegadevice silicon chip, while a small problem is to identify a better interconnect material. Through interactions with the industry that uses research results, real industry problems can be identified.

Effective dissemination may require going an extra step before research is usable by an industry. This may consist of carrying the research to a more advanced state or presenting it in a different form. The SRC has found that joint meetings of university and industry specialists, special short courses to help transfer newly developed technology, and the SRC electronic data base are effective additions to the normal channels of technical communication. Most important, the interaction is not necessarily between peers, as is normally the case among university researchers, but may entail a specialist communicating results to a nonspecialist or to a specialist in a different field. These types of communication are more difficult.

Center management is crucial to the ERC's success. In the past a major failing of research centers has been that a center became too dependent on the personal attributes of an individual director; if he decided to go somewhere else, the center ceased operation. The director must have the respect of the center's investigators and sufficient authority to focus the research on the defined goal; yet a management structure that provides continuity and stability must also be developed.

EXPECTATIONS OF A GROWTH INDUSTRY

The integrated circuit industry is a growth industry. Often an industry downturn such as we are now experiencing really means that revenues are flat rather than increasing 20 percent a year. An unfortunate attribute of a modern high-technology growth industry is that there is ample competition both nationally and internationally. Until recently, the United States had a comfortable hold on two-thirds of the world market for semiconductors. Now both the fraction and the comfort level are lower.

A growth industry looks to the Engineering Research Centers for added input and support so as to compete more effectively and to continue growing. The three bases of these expectations are ideas, graduates, and research results. Some discussion of each of these bases is in order.

It has been observed that the best ideas in any field often come from outside the field simply because the internal researchers know too many things that can't be done. This is a simplistic statement, but often true. Excessively specialized knowledge often inhibits innovative thinking. In addition, the nonspecialist often may know of developments in other fields that can be applied to the problems at hand. University-situated Engineering Research Centers have access to a wide variety of knowledge, and their staffs are not exposed to some of the inhibiting constraints found in industry. For these reasons many seminal ideas have originated in university laboratories, and it is natural that we look to the Engineering Research Centers for results. It is important, of course, that Center research address ideas that are applicable to real industry problems.

Graduate students entering industry have spent five years focusing on a given subfield of their discipline. They have performed original research and gained considerable perspective. An industry employer benefits when a graduate's field of concentration is directly applicable to the industry's technology base. Through the employment of the graduate, technology is transferred from university research to the industry. If the field of concentration is not a directly related field, little if any technology transfer occurs. Although graduates can and do change areas of specialization with ease, there is a big difference when a career becomes an extension of university research. Thus the alignment of the Center's research with the needs of a client industry becomes more important. The high value of the graduate to the industry also increases the incentive to involve as many students in the research as possible. As one figure of merit, the SRC has used the ratio of total contract costs to the number of graduate students participating in the research.

Research results are difficult to evaluate because often they pass through many hands before finding final application. The applicability of university research varies widely from field to field. In the field of integrated circuit technology, for example, the software engineering from computer-aided design efforts is often directly applied by industry. This is also true of system architecture, as in the Intel commercialization of the Cal Tech hypercube architecture. In process-related research—such as dry etching, ion implantation, and low-pressure oxidation—it is sometimes difficult to track a given result to its eventual application. As a result, advances in processes and device technology may have many sources. For the Engineering Research Centers it is important to understand the means by which research results will be transferred and applied to particular fields.

In conclusion, the Engineering Research Centers and their universities have the opportunity to make a significant difference in the engineering research world and in their client industries. Strong institutional support and attention are required for these opportunities to be realized. The entire engineering community will be watching with anticipation for the results.

Biotechnology and the Healthcare Industry: Expectations for Engineering Research

STEPHEN W. DREW

The United States holds a commanding lead over other nations in the biological sciences, especially in the area of molecular genetics. Yet while opportunities in the worldwide market for commercial biotechnology are exciting, the United States faces severe engineering limitations that affect its ability to maintain world dominance in this field. Other papers in this volume describe engineering research in fields that enable us to place complex robotics and advanced telecommunications systems in composite-material vehicles that can orbit the earth in 90 minutes. Other research has led to systems that produce solid-state devices capable of processing up to one trillion bits of information in a second, and which, as aggregates, begin to approximate human intelligence. In stark contrast to these striking synthetic creations, we live in a biological world in which the engineering of biochemistries has barely begun.

HEALTHCARE: THE PHARMACEUTICAL INDUSTRY

Domestic expenditures for healthcare in all categories exceeded 10 percent of the gross national product for the first time in 1983. One of the major healthcare industries, pharmaceutical manufacturing, focuses on the discovery and development of drug treatments for the prevention, cure, or moderation of disease states. The U.S. pharmaceutical industry participates in a world market for human and animal healthcare drugs that exceeded $35 billion in 1983, and which is expected to maintain steady growth. The United States is a net exporter of pharmaceuticals, although the rate of net export growth has slowed in recent years in the face of increasing international competition.

Biotechnology associated with new drug discovery, drug design, drug synthesis, and scale-up to manufacturing is an integral part of this industry. Biology has always played a central role in the discovery of new drugs. Furthermore, the engineering of biological systems has played an important role in drug manufacture, accounting for roughly 23 percent of annual sales. Biological routes to new products will become more and more important as we move toward the new century.

The pharmaceutical industry has been the first to feel the impact of a revolution in the biological sciences. The enabling science and evolving technology of genetic recombination have been the most evident aspects of this revolution; they have catalyzed an explosive growth in our knowledge of how disease states evolve, advance, and can be counteracted or prevented. New, more effective discovery screens (testing strategies) for pharmacologically active compounds have been developed, and a wide variety of compounds with the promise of high medical and commercial value have been identified.

Startling advances in molecular biology have spurred the growth of biotechnology, but the insights, opportunities, and challenges are by no means limited to molecular genetics—they reach far beyond the current applications of molecular genetics. A full partner in the intense research in new biology, and fueled by unfolding insights into the mechanisms, advance, and control of disease, the pharmaceutical industry is hurtling toward the future.

ENGINEERING RESEARCH IN BIOTECHNOLOGY

While basic research in the biological sciences has accelerated dramatically in recent years, engineering research in biotechnology has lagged seriously in the United States. The current focus of engineering research on process development and scale-up to manufacturing has kept pace with new product discovery, but the "margin of comfort" between the completion of process development and licensure has dwindled. The trend toward more complex product chemistries, higher product purities, and increased product stability will only exacerbate this problem and make process economics even more uncertain. The challenge to the engineering community is clear: we must increase engineering research in biotechnology to keep pace with the explosive growth in the biological sciences.

Manufacturing

Engineering research in the manufacture of biological products has focused on four major areas: (1) bioreactors, (2) product recovery, (3) process control and optimization, and (4) drug delivery systems.

Through more than 40 years of experience there have evolved techniques for the scale-up of classical submerged fermentations (the vast majority of which are aerobic). The industry currently has substantial fermentor (bioreactor) capacity designed, for the most part, for slow, low-density fermentations. Much of the equipment is more than 20 years old, but continual upgrading has kept it usable. Still, most of this capital equipment ultimately limits process performance because of a poor-to-marginal design capability for mixing viscous fermentation broths or for achieving high oxygen and/or heat transfer rates that would allow faster, higher-density fermentations. Many of the new recombinant DNA microorganisms possess characteristics that could, biologically, support rapid, high-yield fermentations. As new products from such hosts move toward manufacture, major improvements in bioreactor design will be required. Many questions—both new and old—remain unanswered in this mature field of inquiry.

The technique for recovery and purification of low-to-intermediate molecular weight (500 to 5,000 daltons) compounds is also fairly well developed. Nevertheless, there are many opportunities for improvement through a better understanding of the principles of liquid/liquid and solid/liquid separations applied to these delicate drugs. Extraction, crystallization, and chromatography are old friends to the biochemical engineer; but the need for higher purities, lower costs, and minimum environmental impact will demand a level of performance that is not currently available.

The recovery and purification of macromolecular products presents special challenges that are only partially met by today's engineering tools. The activities of biopolymers—whether physical, chemical, or immunological—depend on precise conformation, starting with primary structure and proceeding in many cases through quaternary structure. We know very little about the factors in product recovery that influence macromolecular folding (or misfolding), and even less about the potential for postbiosynthesis restructuring or modification of proteins and other biopolymers. The requirement for high purity is particularly demanding, since nonproduct macromolecules may possess physical and chemical characteristics that are quite similar to the product of choice.

Process control technology in the pharmaceutical industry has kept pace with the advances in the chemical process industries. In most cases the control capabilities for batch operations exceed those in other industries. Yet while process control of bioreactors is rapidly maturing, the directed control of discrete cellular biochemistry in bioreactors is grossly immature. Optimization of microbial processes has proceeded in a largely empirical fashion over the last 40 years. The efforts have been remarkably successful, but the pace of development ultimately limits the potential of bioprocess engineering. Engineering research can help by focusing on the kinetics, thermodynamics, and pathway coordination of microbial processes. Some valiant efforts at bioprocess modeling and structured optimization have been made, but much more fundamental work is needed.

The development of systems for drug delivery has become an important area for engineering in the pharmaceutical industry. Engineering research on the movement of molecules in human and animal systems is required so that more effective ways of maintaining optimum dose, minimizing side effects, and directing drugs to their targets can be found.

Drug Design and Synthesis

The use of aerobic fermentation in the biosynthesis of pharmaceutical agents is well established, and many of the comments above relate to this route to new drugs. The advent of powerful new techniques of genetic recombination makes possible the synthesis of exotic mammalian proteins in simple microbial cells by using classical fermentations. Drug modification by single-step biotransformation (hydroxylation, group elimination, etc.) is well known, if infrequently applied. Unfortunately, the process engineer has taken a predominantly advisory role in the development of these chemistries, in contrast to the leadership role taken in scale-up of the process. The power of biochemical synthesis is too appealing to allow this trend to continue.

The future will see increasing constraints on commercial synthetic chemistry. While chiral synthetic technique breathed new life into the organic synthesis of new drugs, the trend toward more complex chemistries is likely to continue. Goal-oriented engineering research can help to identify biochemistries that can extend the range of more classical organic chemistries. Biochemistries that can spare the use of expensive (and occasionally toxic) solvents and reagents are needed. Biochemistries that function effectively at high substrate concentration in organic solvents or in mixed aqueous/organic systems are needed. Biochemistries that function over a wider range of temperatures are needed.

Opportunities in the engineering of process biochemistries abound, not just in the commercial-scale synthesis of drugs, but in drug discovery and drug design as well. If any criticism can be leveled against the biochemical engineering community it is that we are not sufficiently in tune with the chemical potential of biotechnology. We need to focus more intensely on the application of process biochemistry.

New Drug Discovery

The traditional challenge of biochemical engineering has been to scale-up a single process to large volumes for commercial manufacture. The engineering challenge in new drug discovery is to scale-up discovery screens designed for small-volume samples so they can handle a very large number of samples. The scientific bases for new drug discovery have been developing at a terrific rate in recent years. Insights from mode-of-action studies, identification of factors in mammalian biochemistry, structure-activity analyses, and other kinds of knowledge have supported the invention of a myriad of novel, highly selective screens for new drugs. By contrast, the engineering research aspects of new drug screening are virtually undeveloped. Fundamental problems in heat, mass, and momentum transport, in the kinetics of receptor-site assays, and in the micromanipulation of samples and reagents await resolution. Challenges in the uniform cultivation and preservation of an extremely wide variety of microorganisms at very small scale await engineering analysis. A quantum jump in the discovery of new, life-saving, natural product drugs will require more than the new biology; it will require engineering research.

CONCLUSION: THE NEED FOR ENGINEERS AT THE INTERFACE

The life science interface with engineering is no longer latent. Clearly the next generation of bioengineers must have a broader and deeper knowledge of the life sciences. At the moment biocatalysis, biochemistry, microbiology, and molecular biology are a few of the areas needing particular focus by engineering students. The educational opportunities in bioengineering systems research are immensely important in this regard. The Engineering Research Centers will bring together engineers and life scientists in an environment primed for discovery.

The study of the application of science is easy
to anyone who is master of the theory of it.
—Louis Pasteur

Challenges for Government

NAM P. SUH

Several papers in this volume discuss engineering issues in the context of the year 2000. The future is very pertinent to the Engineering Research Centers (ERCs) and the National Science Foundation. The NSF is one of the few investment organizations we have within the federal government. The returns on investments we make today usually are not realized for 10 to 20 years, which means we are looking at the year 2000 when we talk about current NSF programs.

In that context, much of what these papers have to say is relevant and thought-provoking. Essentially, what they discuss can be grouped in two categories: problems and opportunities. The problems and opportunities Professor Quinn writes of are very much to the point. He says that we will have a major increase in the world's population; and that, in turn, has a number of implications. He notes that the gross national product of China has grown at an annual rate of 7 or 8 percent over the past several years. If we extrapolate that rate of growth to the year 2000, China's standard of living will be quite high. When that happens, China's natural resource requirements will be larger. If the Chinese standard of living reaches even 10 percent of ours, China's need for materials and other natural resources is going to be about 30 percent of that of the United States, since the population of China is four or five times as large as ours. In order to deal with this essentially global problem, we must begin serious inquiries into the more effective utilization of materials and energy through creative fundamental research. Are our institutions ready to deal with these major issues of mankind? In spite of the fact that we are facing major problems in the world, our educational institutions are not producing enough people who can deal with the large systems issues involved.

Where there are problems, there are also opportunities—opportunities to improve information technology, to create new manufacturing technologies, and to foster emerging technologies such as biotechnology, to name a few. We even have opportunities in critical technologies and in such mature industries as steel. A major goal of the ERC program is to link problems with opportunities. In this linking process all of us have a role to play.

The ERCs must choose important engineering problems that require a cross-disciplinary approach, and provide solutions and manpower. The ERCs and their industrial partners must identify the problems that hold the key to our future technological progress. The role of government is not to dictate what the community should work on and what it thinks are the important problems. Instead, it must rely on the community to develop a consensus about the areas requiring research emphasis.

However, government does have an important role to play. The role of NSF is that of a catalyst. It is an enabling agent that helps the universities to accomplish their goals. It is also a facilitator: it can make the collaboration between the universities and industry easier. Indeed, it can help the university people to fulfill their dreams for excellence in higher education. In addition to these roles, the NSF must protect and promote the public interest. The ERC program enables the NSF to fulfill all of these roles.

In his paper Dr. Hall states that the NSF is not going to micromanage the ERCs. NSF's policy is formulated in the spirit of the role of catalyst: we would like to promote the goals of the ERCs, but we would like to let the ERCs decide what they ought to do by letting university people and industrial people jointly establish their common agenda.

NSF's strategic plan for the ERCs consists of the following elements. First, we would like to establish between 20 and 25 Centers during the next two or three years. Next year we are planning to establish six Centers, if our budget wins the support of Congress. If not, we may have to decrease the number of new starts. Second, NSF plans to establish management teams for the ERCs within the NSF, and to render assistance to the ERCs to ensure their success. We will do whatever we can to help, and we will provide the Centers with whatever they need to achieve their goals. Third, NSF plans to secure for the ERC program the support of Congress, the Office of Science and Technology Policy, the Office of Management and Budget, the National Science Board, and the engineering community at large. I will be spending a great deal of time trying to articulate the need for this type of Center. Finally, as the funding agency, the NSF plans to monitor the progress of these Centers and to make sure they carry out the goals set forth in their proposals.

The NSF is also working to find ways for state governments to fund some of the ERCs within their own states. Once the state governments establish the infrastructure for research at their state-supported institutions, it will be easier for those schools to acquire NSF funding, since they will be more competitive.

In addition to these plans we have a number of other complementary programs within the NSF Engineering Directorate. We have been supporting individual researchers through single-project programs, in which we support one researcher or a group of researchers. This kind of grant may also be used to establish or upgrade the academic research infrastructure. For example, if a university is interested in establishing a biotechnology program, it does not have to rely solely on the ERC program. We have a research program for biotechnology which is designed to help universities in establishing their academic infrastructures. We also have very successful programs that have promoted cooperation between industry and universities—i.e., the Industry/University Cooperative (IUC) Research Programs and the IUC Center Programs. These programs have established a large number of successful cooperative research centers in the past. We must strengthen these programs in the years to come.

The NSF is planning new initiatives for FY 1987. The new programs deal with engineering manpower, facilities, access to federal and national laboratories, and generic engineering systems. In developing these plans we need the ideas and counsel of the engineering community to ensure that the new initiatives are executed in a most effective and rational way.

We hope that the Engineering Research Centers established so far will become role models for successful ERCs. Other institutions can then emulate them and develop equally successful ERCs in the years to come. However, we are realists. We don't expect that every one of these Centers will be successful. But if only a few of them succeed we can use them as role models in establishing new ones. We have a great deal to learn. If some Centers fail, stones should not be thrown at the whole concept.

In the final analysis, no government can be greater than the people it represents—especially with the form of government that we have. Continuing support for the ERC concept will be essential to the continuing support of ERCs. With the support of the entire engineering community behind the ERCs, I think Congress will continue to look favorably upon this endeavor in the years to come.

Implications and Challenges for Industry

JAMES F. LARDNER

The recommendations of the National Academy of Engineering to the National Science Foundation (NSF) about establishing Engineering Research Centers reflect the concern of many business and academic leaders that U.S. engineering education today does not meet industry's real needs. I believe that much of the blame for this situation lies with industry (although academe has too often been a willing and active contributor). In accepting, without complaint or comment, the conventional products of U.S. engineering education; in helping create shortages of qualified engineering faculty by hiring talented faculty members away from teaching; and, in ignoring the dearth of adequate research into manufacturing itself, industry has contributed to the problem it has finally identified and would like to see corrected.

Why is it that industry apparently has acted against what clearly were its own best interests? I suggest the reason is found in the essence of traditional U.S. manufacturing culture. During most of our national industrial development, American manufacturing companies enthusiastically embraced the principles of specialization and division of labor to address the increasing complexity of products and of the manufacturing environment. For a long time this approach worked.

As the techniques of specialization and division of labor were refined, manufacturing became increasingly efficient. Ideas and materials were transformed into products using fewer resources per unit of output. Productivity increased, and with it the wealth of the nation. At the turn of the century this view of industrial organization was dignified by Frederick W. Taylor with the term "scientific management." Unfortunately, this approach to dealing with complexity turned out to be neither very scientific nor very good management, but that fact was not recognized for another 70 years.

What the division of labor and specialization finally caused was the "dis-integration" of manufacturing. Continued growth in the complexity of products, processes, and the environments of manufacturing operations all led to additional specialization and to greater and greater division of responsibility. Unfortunately, we have only now begun to recognize that the solution we adopted with such confidence has resulted in inefficient, unresponsive organizations that are difficult to manage, resistant to change, slow to adopt new technologies, and suffering from formidable communication problems. These negative and unexpected results have caused thoughtful industrial managers to consider reintegrating manufacturing so as to survive in an intensely competitive world. (I hope it is by now agreed that manufacturing spans the range of activities from product concept and design to support of the product in the field.)

There is a powerful case to support the conclusion that the organizational culture in a large part of U.S. industry has caused too many American companies to be late in identifying needed changes in manufacturing management, and late in educating manufacturing management to use resources effectively enough to survive in international competition. Growing recognition of the cause and nature of this problem has led some perceptive individuals to argue for significant changes in the way we educate engineers. These recommendations have been eloquent and forceful. The question is, are they valid? Should we seriously modify the way we educate engineers?

The answer, I think, is "yes and no." "Yes" for some engineering students, but "no" for the rest. Many of us who helped develop recommendations for establishing the NSF's Engineering Research Centers feel strongly that a solid foundation in engineering fundamentals remains an essential part of a quality engineering education. We also think that the Centers can help fill a critical void in engineering education for some engineering students. The Centers can become a unique and major factor in advancing the concept of manufacturing as a science. Important features of successful Centers would be multidisciplinary research, substantial industry involvement in identifying areas for research, industry support for projects selected, and development of a codified body of new knowledge and instructional material about manufacturing and manufacturing problems. This should create an environment in which the problems and benefits of integration can be studied, and where the lessons from past failures can be learned. Clearly, industry has a vital interest in supporting these initiatives.

However, the challenge for industry goes beyond simply supporting the

Engineering Research Centers financially if the Centers are to achieve their objectives.

1. Industry must help define the environments for valid manufacturing research. Most engineering campuses have had difficulties in attempting to create realistic manufacturing environments to challenge both students and faculty.

2. Industry must help identify and define manufacturing research needs that offer intellectual challenges to the academic community, that are commensurate with established research activities on university campuses, and that will withstand the scrutiny of peer review. In the past, given the emphasis on specialization and division of labor, industry was generally content to accept and support research projects selected and defined by a principal investigator. Now, as industry struggles with the task of reintegration, problems increasingly are seen as multidisciplinary, and the lack of research to help solve them is of growing concern. Industry has a responsibility to make this concern known and understood.

3. Industry must recognize the need to support university programs to recruit and retain adequate numbers of qualified engineering faculty. Without sufficient qualified, motivated faculty, the Centers cannot succeed.

4. Industry should be prepared to support the development and publication of instructional material based on manufacturing research findings. The apprenticeship method of teaching engineers about manufacturing simply isn't sufficiently rapid, nor is it as effective as it needs to be if we are going to change our manufacturing culture to survive new global competition.

5. Industry must find ways to provide real-world situations for conducting research, and to make available selected, experienced industry representatives for research projects.

6. Industry must provide constructive input into program evaluation in order to enhance the contributions of research findings and of the graduates the Centers produce.

7. To contribute to the success of the enterprise, industry must recognize, hire, and reward graduates of the Engineering Research Centers, offering opportunities commensurate to the potential these individuals have.

These will be new and difficult challenges for industry. It has not been a hallmark of U.S. industry to look to academic research for help with problems as fundamental and broad as the reintegration of manufacturing, or for insights into how manufacturing organizations might be reorganized to make this reintegration possible. Industry has not traditionally turned to engineering schools for help in managing manufacturing, but there is increasing evidence that in evaluating the changes considered schools may be the preferred resource.

Finally, it is important to remember that industry and academe operate by different time scales. Everyone involved in the ERC effort knows it will be some time before the products of the Centers—whether graduates or research findings—will be available to industry, and even longer before these products will have measurable impact on industry results.

For the interim, industry will have to "wing it," to depend on experience, common sense, and intuition to steer an uncharted course. Despite the absence of immediately useful output applicable to industry problems, management needs to maintain a belief in and provide support for the ERC concept until the first results can be evaluated.

Today's situation reminds me of a time in my naval career when I was "in destroyers," operating with a carrier task force. I don't know how they do it today, but back then when we changed the fleet axis, the destroyers would race through the maneuvering ships at high speed on an approximate course, chosen to avoid collisions, to get close to their new screen stations. Only as they approached their new stations did the fine maneuvering begin. Varying course and speed slightly but continuously, if successful they dropped in, right on station, exactly where they belonged, and their captains lost no promotion numbers.

I think industry today faces a similar situation. We are changing from where we were to where we have to be, and we have no time to spare. As we move closer to where we want to be, we will require special skills and knowledge that can put us right on station. I think these can come— to an important degree—from the Engineering Research Centers, and I believe these Centers deserve industry support.

Challenges for Academe

H. GUYFORD STEVER

I have the last word in this volume; but those universities that will host the Engineering Research Centers (ERCs) will have the last word on whether the Centers are successful.

In these pages many leaders of American industry, government, and academe discuss how important the Centers are to the nation's future. I think it is the concept itself that is most important—that of pooling our engineering research efforts on a bigger and broader scale. Teams of engineers and scientists from many disciplines, from both academe and industry, working together, with the cooperation and support of government, to target problems of importance to our competitive future—that is an exciting idea.

It is not a new idea, of course. It has been tried before, but usually on a smaller scale and with less clarity of purpose, less sense of urgency. However, there is often a great gulf between ideas and reality.

The message running through these papers is: The Centers are needed, and we must make them work! But those in academe, especially, know well what the real problems will be. Larry Sumney suggests some of them. Young faculty members will be wary, maybe reluctant to participate because of their fears about unknown (or perhaps too well known) threats to their careers. Cross-disciplinary research is usually not an accepted route to advancement; in many institutions that battle has yet to be fought.

When I was the newly appointed president of Carnegie-Mellon University, a group of professors came in to see me. These were distinguished professors from different departments who wanted to start what we called then an interdisciplinary center. I listened to them. Their ideas were just great, and I was quite excited about it; but at the end of the presentation they said, "Now, we have to find someone whom we can bring to the university to lead this center." I said, "Stop right there. When any of you strong people in your disciplines who have all these good ideas are willing to risk your career to lead this effort, then I will go along with it." About a year later, two of them came in and took the responsibility. They changed their careers. I think they are very happy today that they did, but the fact is that they took a risk. Graduate students are also going to have to take a risk. Many of them may be quite excited when they notice all the drum-beating that has accompanied the ERC program. But some will look at the situation and conclude that the disciplinary approach to education is still very strong.

Existing departments may not readily accept the ERCs. The resistance may not surface until the going gets tough for one reason or another; but retrenchment into the disciplinary fold has always been the instinctive response in such circumstances.

Another problem is what happens if, after the seed funds are withdrawn and the ERC has becomes self-supporting, the Center encounters a downturn in the nation's economy. Industry funding may diminish. What happens to the ERC then? Will it be a case of "last to arrive, first to leave"?

What can we do to make the world safe for ERCs?

Changes will have to occur, of which the first will be a change in "campus sociology." As James Lardner's paper points out, some industries are already wrestling hard with this requirement in their own context. They can't avoid it—their improved performance demands this adaptation.

But universities have so far not accepted the proposed mode. The disciplinary structure has remained essentially intact, preferring instead to split off new disciplines to accommodate the explosion of knowledge and the emergence of problems such as the environment, or new technologies such as the computer. That approach is no longer completely sufficient. Of course the disciplines must continue to be strong. But, as we are already seeing in efforts such as MIT's new Interdepartmental Biotechnology Program, the cross-disciplinary approach must increasingly be reflected in the organizational structure of science and engineering.

Schools must figure out a way to accomplish research goals of a cross-disciplinary nature while still maintaining strong disciplinary depth. The reward system will have to be modified to accommodate this requirement. That is a challenge that every school will have to address in terms of its own particular situation, its own "culture." If the schools fall short of that, in Larry Sumney's words, "we will all lose."

According to the National Science Foundation's program announcement for FY 1986, one of the four criteria upon which the next round of ERC proposals will be judged relates to this very thing: a concern for the "effect of the research on the infrastructure of science and engineering." Any proposal demonstrating a commitment to this kind of change is likely to be a stronger proposal, in the eyes of NSF.

Second, schools will have to alter their relations with the outside world. Faculty consulting and small-scale cooperative research with industry are fine, and should continue. But they are not enough. Universities will have to open their doors in new ways, defining strategies for making and cementing ties with state and local governments, other schools, and companies large and small. These ties should be stable, long-term, and mutually beneficial.

Third, and perhaps most fundamental, a sensitivity must emerge in the university community regarding the needs of the nation, regarding the situation of the nation with respect to economic and competitive fortunes to which engineering holds a very important key. The Engineering Research Centers are being created to improve our national technological productivity and competitiveness. This can only be done through a systems approach to real-world problems—not through abstraction and analysis for its own sake. A new generation of engineering students has to be educated to think and function in the cross-disciplinary context.

I think the ultimate challenge in all this lies with the individual, as it always does when change must take place. As I have pointed out, the young faculty members who work in the ERC programs will have to be courageous people. They will have to be committed to goals and methods that the power structure may not share, that even many of their academic peers do not share.

Graduate and postgraduate students who participate in the ERCs will also need to have commitment. When they have finished their education they will have a major decision to make: whether to go into industry or to join a faculty. The latter choice may be the only avenue by which real change can be brought to the disciplinary structure, since those individuals will have come up through the new system.

Academic administrators who want the ERCs to succeed will have to have the commitment necessary to push against disciplinary barriers and to protect the ERCs from adverse pressures.

Industry managers will have to be ready to be committed to the success of the program, even when continued support is painful to the company. They may have to convince boards of directors and, ultimately, stockholders, and persuade them to share that commitment.

In many cases, the academic institution has to make a commitment to individuals if they are going to take the risk of participating. Some will participate no matter what, because they share a conviction about what these Centers represent. Yet their fate is in the hands of people who will be difficult to persuade of that vision, that commitment. This is where we can clearly see the fragility and the vulnerability of the fledgling ERCs.

By funding these six Centers, the National Science Foundation has taken the first strengthening steps toward a new approach to engineering research, education, and practice. I am willing to bet very strongly that the initial impetus has been and will continue to be very well received by the Congress and the administration. I cannot conceive of an administration that would resist this kind of approach now or sometime in the future, and therefore I think it is on very good ground. But the battle is by no means over for the ERCs. In wartime the tank units have to select a point tank for every one of their advances, and it can be imagined what chances that point tank has to take. It is the same with the ERCs. We have only a few point units out there, and we had better make sure that they are very well supported by everyone concerned. Eventually we will have a larger number of units, and then we can sit back and let them compete in a rough-and-tumble world. But we had better make it a good world for them for a while.

Biographies

DAVID C. ANDERSON, Professor of Mechanical Engineering, is Director

of Purdue's Computer-Aided Design and Graphics
Laboratory, a facility dedicated to the study of computer-aided design and computer-aided manufacturing
(CAD/CAM). Dr. Anderson is the author of many
articles, and has worked as an industrial consultant in
the area of CAD/CAM. He received his Bachelor's,
Master's, and Ph.D. degrees in mechanical engineering from Purdue.

JOHN S. BARAS is Professor of Electrical Engineering at the University

of Maryland, and Director of that university's newly
established Systems Research Center. He performed
his Master's and Doctoral work (1973) at Harvard
University, in applied mathematics; he also holds a
Bachelor's degree in electrical engineering. Dr. Baras
is a Fellow of the IEEE, and the recipient of numerous
awards for research. The primary focus of his research
has been in control and systems theory.

MOSHE M. BARASH is a Professor of Manufacturing in Purdue's School of Industrial Engineering. Dr. Barash has had extensive experience in industry and nonuniversity research on design of complex machines, instruments, control systems, and production processes and tools. He received his B.Sc. and Dipl.-Ing. degrees in mechanical and electrical engineering from the Technion-Israel Institute of Technology, and from 1947 to 1955 was involved in machine design and research in mechanical systems and instruments (1953) from the University of Manchester, England, where he taught the subject until 1963, when he joined the faculty at Purdue. Dr. Barash has published more than 70 research papers and more than 300 technical articles. He is a Fellow of the American Society of Mechanical Engineers, and was given the Blackall Award by that society in 1983 for the best paper in machine tool technology.

ARDEN L. BEMENT, JR., is Vice-President of Technical Resources for TRW. He received his Ph.D. in metallurgical engineering at the University of Michigan in 1963. He worked in industry initially, including 10 years with General Electric and 5 years in nuclear materials research at the Battelle Memorial Institute; later he was a Professor of Nuclear Engineering (materials) at MIT. From 1976 to 1979 Dr. Bement served as Director of the Office of Materials Science at the Defense Advanced Research Projects Agency (DARPA); he was later Deputy Under Secretary of Defense for Research and Engineering. He is a member of the National Academy of Engineering.

ERICH BLOCH is Director of the National Science Foundation. He joined IBM Corporation in 1952 after receiving a B.S. in electrical engineering at the University of Buffalo (now SUNY Buffalo). He was instrumental in development of the IBM 360 computer (among other projects), for which he was awarded the National Medal of Technology in February 1985. Before coming to NSF, Mr. Bloch was Chairman of the Semiconductor Research Corporation, and served as Vice-President for Technical Personnel Development at IBM from 1981 to 1984. He is a member of the National Academy of Engineering.

W. DALE COMPTON is Vice-President for Research at the Ford Motor Company. He received his Ph.D. in physics from the University of Illinois in 1955. After working at the Naval Research Laboratory, Dr. Compton taught physics at the University of Illinois at Urbana. He joined Ford Motor Company in 1970 as Director of Chemical and Physical Science. He is a member of the National Academy of Engineering, where he served as chairman of the Academy committee that drafted the guidelines for the Engineering Research Centers.

STEPHEN W. DREW is Director of Biochemical Engineering at Merck and Company. He received his Ph.D. in biochemical engineering from MIT in 1974. He was a Professor of Chemical Engineering at Virginia Polytechnic Institute and State University (VPI) before joining Merck in 1980. Dr. Drew is currently a member of the Panel on Bioengineering Systems Research of the National Research Council's Engineering Research Board.

KING-SUN FU, principal investigator for Purdue University's new Center for Intelligent Manufacturing Systems, is Goss Distinguished Professor of Engineering in Purdue's School of Electrical Engineering. He is internationally recognized as a pioneer in the engineering disciplines of pattern recognition, image processing, and machine (artificial) intelligence. Dr. Fu has received numerous honors and awards for his contributions in these areas, and was elected a member of the National Academy of Engineering in 1976. He is the author of four books and numerous book chapters, journal articles, and technical papers in his field. Dr. Fu received a Bachelor's degree in engineering from the National Taiwan University, a Master's degree from the University of Toronto, and a Ph.D. degree in engineering (1959) from the University of Illinois. [Editor's note: Dr. Fu died on April 29, 1985, while attending the symposium on the Engineering Research Centers.]

SUSAN HACKWOOD is Professor of Electrical and Computer Engineering

at the University of California, Santa Barbara; she will be the Director of UCSB's newly established Center for Robotic Systems in Microelectronics. Dr. Hackwood obtained her Ph.D. in solid-state electrochemistry at Leicester Polytechnic Institute, U.K. After completing the Doctorate, she joined AT&T Bell Laboratories, where she remained until 1984. At Bell Labs she carried out a range of research in robotics, and was named Head of the Robotics Technology Research Department.

JERRIER A. HADDAD is a consultant to the National Research Council

and was recently Chairman of its Committee on the Education and Utilization of the Engineer. He joined the IBM Corporation after receiving a Bachelor's degree in electrical engineering at Cornell University in 1945. At IBM he held a number of technical managerial positions; he was IBM Vice-President for Engineering, Programming, and Technology (1967–1977) and for Technical Personnel Development (1977–1981). Mr. Haddad is a trustee of Clarkson College and Chairman of the Engineering College Advisory Council of Cornell University. He has received two honorary Doctor of Science degrees, as well as numerous awards and patents. Mr. Haddad is a member of the National Academy of Engineering and a Fellow of the Institute of Electrical and Electronics Engineers.

CARL W. HALL is Deputy Assistant Director of Engineering for the

National Science Foundation. He has degrees in agricultural and mechanical engineering, and received his Ph.D. from Michigan State University in 1952. He was Chairman of the Agricultural Engineering Department at Michigan State University, and then Dean of the College of Engineering at Washington State University, where he also taught mechanical engineering. Dr. Hall has been active as a consultant in numerous international projects. He has authored many books on energy and food engineering, and is editor of an international journal on drying. He has received a number of awards for his achievements. Dr. Hall is a Fellow of the American Society of Agricultural Engineers, a Life Fellow of the American Society of Mechanical Engineers, and a Fellow of the American Association for the Advancement of Science.

GEORGE A. KEYWORTH II is Director of the Office of Science and Technology Policy and Science Adviser to the President. He received his Bachelor's degree in physics from Yale University in 1963 and his Ph.D. in nuclear physics from Duke University in 1968. He holds honorary Doctor of Science and Doctor of Engineering degrees as well. Dr. Keyworth was associated with the Los Alamos National Laboratory from 1968 to 1981, where he held Division Leader posts in three areas of physics research. He was recently a member of the President's Commission on Industrial Competitiveness.

JAMES F. LARDNER is Vice-President, Component Group, for Deere and Company. After earning a Bachelor's degree in mechanical engineering at Cornell University, he held a number of engineering and manufacturing management positions in Deere's domestic and overseas divisions. In 1980 he was named Vice-President in charge of Manufacturing Development, a position in which he was responsible for the strategic planning and evaluation of new and advanced manufacturing systems and technologies. In his most recent assignment he is responsible for the design and manufacture of the major components which make up John Deere end products.

LEWIS G. ("PETE") MAYFIELD is Head of the Office of Cross-Disciplinary Research (the office responsible for the Engineering Research Centers program) within the National Science Foundation. He received his M.S. in chemical engineering at Montana State College in 1950. After a career in industry and academe, he joined the NSF in 1962, directing programs and divisions concerned with advanced technology applications, integrated basic research, and chemical and process engineering.

R. BYRON PIPES is Professor of Mechanical and Aerospace Engineering at the University of Delaware, and Director of that university's Center for Composite Materials. He received his Ph.D. at the University of Texas in 1972. Since 1977 Dr. Pipes has been associated, first as Acting Director and then as Director, with a Center for Composite Materials at the University of Delaware, which will now be expanded to focus on cross-disciplinary research and education in composites manufacturing. He has authored a number of books and papers on composite materials, and is currently a member of the Panel on Materials Systems Research of the National Research Council's Engineering Research Board. On July 1, 1985, Dr. Pipes became Dean of the College of Engineering of the University of Delaware. Professor Roy L. McCullough, Associate Director, has assumed the duties of Acting Director of the new Center.

JAMES BRIAN QUINN is William and Josephine Buchanan Professor of Management at the Amos Tuck School of Business Administration, Dartmouth College. He received a B.S. in engineering at Yale, an M.B.A. at Harvard, and earned his Doctorate at Columbia University in 1958. Dr. Quinn has taught on the faculty at Dartmouth since 1957. In addition, he has been president of a research, planning, and development consulting firm since 1961 and chairman of high technology start-up firms.

ROLAND W. SCHMITT is Chairman of the National Science Board (governing body of the National Science Foundation), and is also Senior Vice-President for Corporate Research and Development of the General Electric Company. He received his Doctorate from Rice University in 1951, and has been with General Electric since that year. Dr. Schmitt is on the board of directors of a number of nonprofit organizations devoted to science, technology, and medicine. He is also a member of the Council of the National Academy of Engineering.

MISCHA SCHWARTZ is Professor of Electrical Engineering and Computer Science at Columbia University, where he will direct the newly established Engineering Center for Telecommunications Research. After earning a Master's degree in electrical engineering, he received his Ph.D. in applied physics from Harvard University in 1951. Dr. Schwartz was an engineer with the Sperry Gyroscope Company and Professor of Electrical Engineering at the Polytechnic Institute of Brooklyn before coming to Columbia. He is the author of numerous books and publications, and received the IEEE Education Medal in 1983. Dr. Schwartz was nominated to the IEEE Centennial Hall of Fame in 1984. His primary research interests are in communication theory and systems, digital communications, and computer communications. He is currently President of the IEEE Communications Society.

JAMES J. SOLBERG is a Professor of Industrial Engineering at Purdue University and Associate Director of the Computer Integrated Design, Manufacturing, and Automation Center (CIDMAC). He has won numerous awards for teaching and research. Since 1975 Dr. Solberg has conducted research on the mathematical modeling of manufacturing systems. He developed a program called CAN-Q, which is now widely used by industries and universities around the nation. Dr. Solberg received his Bachelor's degree in mathematics from Harvard University, and Master's degrees in mathematics and industrial engineering and a Ph.D. in industrial engineering from the University of Michigan. He joined Purdue in 1971 after three years at the University of Toledo.

H. GUYFORD STEVER is a former Director of the National Science Foundation, and is currently President of the Universities Research Association. He received his Doctorate in physics from the California Institute of Technology in 1941, and holds numerous honorary degrees. He was Professor of Aeronautics and Astronautics at MIT from 1946 to 1965, President of Carnegie-Mellon University from 1965 to 1972, and Director of the Office of Science and Technology Policy and Science Adviser to the President during the Ford administration. Dr. Stever is a director of several corporations, and has been the recipient of numerous awards for his public service. He is a member of the National Academy of Engineering and the National Academy of Sciences.

NAM P. SUH is Assistant Director for Engineering of the National Science Foundation. He performed his undergraduate work in mechanical engineering at MIT, and received the Ph.D. from Carnegie-Mellon University in 1964. Before coming to NSF, Dr. Suh was Professor of Mechanical Engineering at MIT, and Director of the Laboratory for Manufacturing and Productivity there. He has been a director of several corporations involved in technology development; and he is the author or editor of a number of fundamental textbooks in engineering sciences.

LARRY W. SUMNEY is President of the Semiconductor Research Corporation. He received a Bachelor's degree in physics from Washington and Jefferson College in 1962, and a Master's degree in systems engineering from George Washington University in 1969. From 1962 to 1972 he worked at the Naval Research Laboratory as a research physicist and, later, as an electronics engineer. He then joined the Naval Electronics Systems Command, becoming Head of the Solid State and Special Device Technology Branch and, ultimately, Research Director. On assignment to the Office of the Under Secretary of Defense for Research and Engineering, Mr. Sumney managed the formation of the Very High Speed Integrated Circuits (VHSIC) Program, and subsequently became its first director. He joined the newly formed SRC in 1982 as its first executive director.

ERIC A. WALKER is Cochairman of the Advisory Panel for the Engineering Research Centers. He was educated at Harvard University, where he received his Sc.D. degree in 1935; he holds a number of honorary doctorates as well. Dr. Walker taught electrical engineering at the University of Connecticut and at Pennsylvania State University, where he was Dean of Engineering and later President of the University (1956–1970). After leaving Penn State, he was Vice-President for Science and Technology at the Aluminum Company of America (ALCOA). Dr. Walker is a past Chairman of the National Science Board, a past Chairman of the Naval Research Advisory Commission, and was the second president of the National Academy of Engineering. He is currently Chairman of the Board of the Institute for Defense Analyses.

DANIEL I. C. WANG is Professor of Chemical and Biochemical Engineering at MIT, where he will direct the newly established Center on Biotechnology Process Engineering. Dr. Wang holds an M.S. in biochemical engineering from MIT and a Ph.D. in chemical engineering from the University of Pennsylvania (1963). He came to MIT in 1965 after two years as a process development engineer with the U.S. Army Biological Laboratories. He has authored three books, in addition to numerous other publications. Dr. Wang's primary research interests are in the molecular biology of animal cells, bioreactor design and operations, downstream processing, and biochemical process systems engineering.